13-99

88084

088084

D0418261

84507

9. FEB

13. JAN. 19

JAN. 199
RETURNED

0. JUL. 199
RETURNED

-8 DEC. 19

-5 JAN. 199
RETURNED
3. FEB. 199
RETURNED

Don Gresswell L

88084

THE CONSTRUCTION OF BUILDINGS
VOLUME 4

FOURTH EDITION

By the same author

Volume 1
Foundations and Oversite Concrete –
Walls – Floors – Roofs

Volume 2
Windows – Doors – Fires –
Stairs – Finishes

Volume 3
Single Storey Frames – Shells –
Lightweight Coverings

Volume 5
Supply and Discharge Services

THE CONSTRUCTION OF BUILDINGS

VOLUME 4

MULTI-STOREY BUILDINGS, FOUNDATIONS AND SUBSTRUCTURES, STRUCTURAL STEEL FRAMES, FLOORS AND ROOFS, CONCRETE, CONCRETE STRUCTURAL FRAMES, EXTERNAL WALLS AND CLADDING OF FRAMED BUILDINGS

R. BARRY

Architect

FOURTH EDITION

Blackwell
Science

ST HELENS
COLLEGE

690

BAR

101301

LIBRARY

© R. Barry, 1966, 1971, 1986, 1996

Blackwell Science Ltd
Editorial Offices:
Osney Mead, Oxford OX2 0EL
25 John Street, London WC1N 2BL
23 Ainslie Place, Edinburgh EH3 6AJ
238 Main Street, Cambridge
    Massachusetts 02142, USA
54 University Street, Carlton
    Victoria 3053, Australia

Other Editorial Offices:
Arnette Blackwell SA
    224, Boulevard Saint Germain
    75007 Paris, France

Blackwell Wissenschafts-Verlag GmbH
    Kurfürstendamm 57
    10707 Berlin, Germany

    Zehetnergasse 6
    A-1140 Wien
    Austria

All rights reserved. No part of this
publication may be reproduced, stored
in a retrieval system, or transmitted,
in any form or by any means,
electronic, mechanical, photocopying,
recording or otherwise, except as
permitted by the UK Copyright,
Designs and Patents Act 1988,
without the prior permission of the
publisher

First published in Great Britain by
    Crosby Lockwood & Son Ltd 1966
Reprinted 1968, 1969
Second edition (metric) 1971
Reprinted by Granada Publishing Ltd in Crosby Lockwood Staples
    1973, 1974, 1976, 1979
Reprinted by Granada Publishing 1980, 1982
Reprinted by Collins Professional and Technical Books 1985
Third edition 1986
Reprinted by BSP Professional Books 1988, 1990
Reprinted by Blackwell Scientific Publications 1992
Fourth edition published by
    Blackwell Science 1996

Set by DP Photosetting, Aylesbury, Bucks
Printed and bound in Great Britain
at the University Press, Cambridge

The Blackwell Science logo is a
trade mark of Blackwell Science Ltd,
registered at the United Kingdom
Trade Marks Registry

DISTRIBUTORS

Marston Book Services Ltd
PO Box 269
Abingdon
Oxon OX14 4YN
(*Orders:* Tel: 01235 465500
            Fax: 01235 465555)

USA
Blackwell Science, Inc.
238 Main Street
Cambridge, MA 02142
(*Orders:* Tel: 800 215-1000
            617 876-7000
            Fax: 617 492-5263)
Canada
Copp Clark, Ltd
2775 Matheson Blvd East
Mississauga, Ontario
Canada, L4W 4P7
(*Orders:* Tel: 800 263-4374
            905 238-6074)

Australia
Blackwell Science Pty Ltd
54 University Street
Carlton, Victoria 3053
(*Orders:* Tel: 03 9347 0300
            Fax: 03 9349 3016)

A catalogue record for this title
is available from the British Library

ISBN 0–632–03911–6 (BSL)
ISBN 0–632–04116–1 (IE)

Library of Congress
Cataloging-in-Publication Data

Barry, R. (Robin)
    The construction of buildings.
    Vol. 4 published: Oxford; Boston:
Blackwell Science.
    Includes indexes.
    Contents: v. 1. Foundations, walls, floors,
roofs (5th ed. rev.)—v. 2. Windows, doors,
fires, stairs, finishes (4th ed.)—v. 3. Single
storey frames, shells, and lightweight
coverings (4th ed.)—v. 4. Multi-storey
buildings, foundations, steel frames, concrete
frames, floors, wall cladding (4th ed.)—v. 5.
Supply and discharge services.
    1. Building.   I. Title.
TH146.B3   1980   690   81-463308
ISBN 0–632–03911–6

ST. HELENS
COLLEGE

690

88084

JUN 1997

LIBRARY

CONTENTS

INTRODUCTION · vii

CHAPTER ONE · FOUNDATIONS AND SUBSTRUCTURES · 1
Functional requirements – Foundations – Substructures

CHAPTER TWO · STRUCTURAL STEEL FRAMES, FLOORS AND ROOFS · 31
Functional requirements – Methods of design – Steel sections – Structural steel frames – Floors and roofs to structural steel frames – Functional requirements – Floor and roof construction – Fire safety

CHAPTER THREE · CONCRETE · 74
Cement – Aggregates – Concrete mixes – Reinforcement – Formwork and falsework – Prestressed concrete – Lightweight concrete – Surface finishes of concrete

CHAPTER FOUR · CONCRETE STRUCTURAL FRAMES · 96
In situ cast frames – Floor construction – Precast reinforced concrete frames – Lift slab construction

CHAPTER FIVE · EXTERNAL WALLS AND CLADDING OF FRAMED BUILDINGS · 116
Functional requirements – External walls and cladding – Solid and cavity walling – Facings applied to solid and cavity wall backing – Cladding panels – Glass fibre reinforced cement cladding panels (GRC) – Glass fibre reinforced polyester cladding (GRP) – Infill wall framing to a structural grid – Glazed wall systems – Glass – Sheet metal wall cladding – Sheet metal wall panels

INDEX · 173

INTRODUCTION

There have been few changes in the basic form of the structural steel, skeleton frame since the first frame was erected in Chicago in 1883. Recent innovations such as the slimfloor and the parallel beam frame, described in this edition, have been adopted to accommodate services in the depth of the floor rather than as a fundamental change in the form of the frame.

The reinforced concrete structural frame, which came into general use in the middle of the twentieth century and was initially used as a substitute for the steel frame, because of the shortage of steel, is to this day much used as a skeleton structural frame. There is little to choose between the reinforced concrete and the skeleton steel frame as regards cost, speed of erection and convenience.

Because of its initial, wet plastic form that facilitates moulding to any shape, reinforced concrete has been, and still is, widely used for precast frames and precast wall frames particularly in those countries where air temperature is, for some months, below freezing.

The use of the initial plasticity of concrete in the construction of unconventional forms has been little exploited other than in such signal buildings as the Sydney Opera House and the shell forms designed by F. Candella in South America.

The appearance of multi-storey buildings has been, and still is, subject to change in the choice and use of materials as wall cladding. Buildings may have the appearance of solid loadbearing structures by the use of natural stone or brick facings fixed to and supported by the frame, be faced with panels of concrete, GRC, GRP, sheet metal or glass fixed to and supported by the frame in a variety of forms.

The so-called high technology buildings of recent years may have some real or false structural members exposed, services such as lifts and pipework exposed on the face of the building and large areas of glass. This artifice is no more high technology than another more conventionally faced building where the services are concealed within the envelope.

What has undergone little change in form is the foundations to multi-storey buildings except that the increasingly conservative approach to loads on foundation bases has meant that the mass of the foundation alone often exceeds that of the structure it supports.

In this edition such innovations as flowdrill jointing to hollow rectangular sections, parallel beam, top hat sections and slimfloor construction have been included. Details of fixings for stone and brick cladding have been revised and expanded and the provisions in The Building Regulations (Amendment) Regulations 1994 have been included.

AUTHOR'S NOTE

For linear measure all measurements are shown in either metres or millimetres. A decimal point is used to distinguish metres and millimetres, the figures to the left of the decimal point being metres and those to the right millimetres. To save needless repetition, the abbreviations 'm' and 'mm' are not used, with one exception. The exception to this system is where there are at present only metric equivalents in decimal fractions of a millimetre. Here the decimal point is used to distinguish millimetres from fractions of a millimetre, the figures to the left of the decimal point being millimetres and those to the right being fractions of a millimetre. In such cases the abbreviations 'mm' will follow the figures e.g. 302.2 mm.

CHAPTER ONE

FOUNDATIONS AND SUBSTRUCTURES

The foundation of a building is that part of the substructure which is in direct contact with and transmits loads to the ground. The substructure is that part of a building or structure which is below natural or artificial ground level and which supports the superstructure. In practice the concrete base of walls, piers and columns and raft and pile foundations are described as the foundation.

The foundation of a building is designed to transmit loads to the ground so that any movements of the foundation are limited and will not adversely affect the functional requirements of the building. Movement of the foundations may be caused either by the load of the building on the ground or by movements of the ground independent of the load.

Ground movements, due to the applied load of buildings on foundations, cause settlement by the compression of soil below foundations or because of shear failure due to overloading.

Settlement movements on non-cohesive soils, such as gravel and sand, take place as the building is erected and this settlement is described as 'immediate settlement'. On cohesive soils such as clay, the settlement is gradual as water or water and air are expelled from pores in the soil. This settlement, which is described as 'consolidation settlement', may continue for several years after completion of the building.

Movement of the foundation by settlement is limited, by the design of the foundation, to avoid damage to connected services and drains and to limit relative movement between different parts of the foundation which otherwise might cause distortion of the structure and damage to finishes, cladding or structural members.

Movements of the foundation that are independent of the applied loads of buildings are due to seasonal changes or the effects of vegetation that lead to shrinking or swelling of clay soils, frost heave, changes in ground water level and changes in the ground due to natural or artificial causes.

FUNCTIONAL REQUIREMENTS

The functional requirements of foundations are:

Strength and stability

The requirements from Part A of Schedule 1 to the Building Regulations 1991, as amended 1994, relevant to foundations are, under the heading 'Loading', that 'the combined dead, imposed and wind loads of the building be safely transmitted to the ground without causing such movement of the ground as will impair the stability of any part of another building' and under the heading 'Ground movement', that 'the building shall be constructed so that ground movement caused by swelling, shrinking or freezing of the subsoil or landslip or subsidence (other than subsidence arising from shrinkage) in so far as the risk can be reasonably foreseen, will not impair the stability of any part of the building'.

The first requirement, under the heading 'Loading', is concerned with the bearing strength of the ground relative to the loads imposed on it by the building. The foundation or foundations should be designed so that the combined loads from the building are spread over an area of the ground capable of sustaining the loads without undue movement. The reference to movement of the ground that might impair the stability of another building is presumably to the pressure on the ground from the foundations of a new building increasing the load on the ground under the foundations of an adjoining building and so increasing the possibility of instability. The swelling, shrinkage or freezing of subsoil is described in Volume 1 and later in this chapter relative to the general classification of soils.

Land instability

The term 'Ground movement', used in the requirements from the Building Regulations relevant to foundations, is more usually expressed as land instability.

Land which is unstable, or may become unstable, is widespread in Great Britain. The extent of unstable land varies from the comparatively large areas affected by coal mining to small areas affected by local quarries. Instability which is caused by natural processes, such as landslip of sloping strata, may be accelerated by human activities such as mining and earth excavation.

Landslip may be broadly grouped under the headings:

- Landslip
- Surface flooding and soil erosion
- Natural caves and fissures
- Mining and quarrying
- Landfill

Landslip

Landslip may occur under natural slopes where weak strata of clay, clay over sand or weak rock strata may slip down the slope, particularly under steep slopes and where water acts as a lubricant to the slip movement. Landslides of superficial strata nearest to the surface, which will be most noticeable and therefore recorded, are those that will in the main cause land instability that may affect the foundations of buildings. Landslides of deeper strata that have occurred or may occur, generally go unnoticed and will only affect deep excavations and foundations.

The most noticeable landslides occur in cliff faces where the continuous erosion of the base of the cliff face by tidal movements of the sea undermines the cliff and causes collapse of the cliff face and subsidence of the supported ground. Similar landslip and subsidence may occur where an excavation is cut into a slope or hillside. The previously supported sloping strata are effectively undermined and may slip towards the excavation. Landslip is also common around excavations for deep coal mining which may break through sloping strata and so encourage landslides.

The Department of the Environment has commissioned studies and prepared reports of areas liable to land instability in and around the coal mining areas of Great Britain. Similar studies have led to reports of areas liable to land instability due to landslip around areas of metal, stone, chalk and limestone quarrying.

Surface flooding

Surface flooding may affect the stability of surface ground and the seasonal movement of water through permeable strata below the surface may cause gradual erosion of soils and permeable rocks that may lead to land instability. The persistent flow of water from fractured water mains and drains may cause gradual erosion of soil and lead to land instability. The incidence of surface flooding and erosion by below surface water is, by and large, known and recorded by the regional water authorities.

Natural caves and fissures

Natural caves and fissures occur generally in areas of Great Britain where soluble rock strata, such as limestone and chalk, have been eroded over time by the natural movement of subterranean water. Where there are caves or small cavities in these areas near the surface, land instability and subsidence may occur. The Department of the Environment has prepared a review of information on the incidence of such cavities in the form of regional reports and maps showing the location and nature of known cavities and the likelihood of land instability due to the cavities.

Mining and quarrying

Mining and quarrying of mineral resources has been carried out for centuries over much of England and parts of Wales and Scotland. The majority of the mines and quarries have by now been abandoned and covered over. From time to time mining shafts collapse and ground, filled over quarries, may subside. There is potential for land instability and subsidence over those areas of Great Britain where mineral extraction has taken place. The Department of the Environment has commissioned surveys and produced reports of those areas known or likely to be subject to land instability due to mining and quarrying activities. There are ten regional reports and atlases indicating the location of areas that may be subject to land instability subsidence. Coal mining areas have been comprehensively surveyed and mapped and reported. Other areas where comparatively extensive quarrying for stone, limestone, chalk and flint has taken place have been surveyed, mapped and reported. Less extensive quarrying, for chalk for example in Norwich, has been included. The reports indicate those areas where subsidence is most likely to occur and the necessary action that should be taken preparatory to building works.

Landfill

Landfill is a general term to include the ground surface which has been raised artificially by the deposit of soil from excavations, backfilling, tipping, refuse disposal and any form of fill which may be poorly compacted, of uncertain composition and density and thus have indeterminate bearing capacity and be classified as unstable land.

Of recent years regional and local authorities have had some control and reasonably comprehensive details of landfill which may give indication of the age, nature and depth of recent fill. The land over much of the area of the older cities and towns in this country, particularly on low lying land, has been raised by excavation, demolition and fill. This over-fill may extend some metres below the surface in and around older settlements and where soil excavated to form docks has been tipped to raise ground levels above flood water levels. Because of the variable and largely unknown nature of this fill, the surface is in effect unstable land and should be considered as such for foundations. There are no records of the extent and nature of this type of fill that has taken place over some considerable time. The only satisfactory method of assessing the suitability of such ground for foundations is by means of trial pits or boreholes to explore and identify the nature and depth of the fill.

Site exploration

As a preliminary to the design of foundations for buildings it is necessary to determine the nature and variability of the soil strata that underlie the building site and assess those properties of the subsoil that may affect the performance of the building.

An inspection of the site, the natural surface of the ground and natural vegetation, evidence of marshy ground, signs of ground water and flooding, irregularities in topography, ground erosion and ditches and flat land near streams and ditches, where there may be soft alluvial soil, will provide an overall indication of the nature of the subsoil. Information on subsoil conditions from county and local authorities, geological surveys, aerial photography, Ordnance Survey maps and works for buildings and services adjacent to the site will provide further reliable information.

To obtain a more precise knowledge of the nature and variability of the subsoil it is necessary to determine the thickness, depth, properties and any major changes in the subsoil strata that are likely to be significantly affected by structural loads.

Trial pits and boreholes

For exploration to shallow depths of up to about two metres, trial pits are excavated by hand over the area of the site likely to be affected by the foundations. The advantage of trial pits is that the sides of the pit can be inspected at all levels.

Where it is necessary to explore to depths greater than two metres, samples of soil are taken from boreholes that are drilled by hand or by means of power operated drills that take out samples of soil at regular intervals. The samples are collected and a note is made of the depth at which the sample was taken and sections of the subsoil strata are prepared. Selected samples of soil are then tested to determine grading of particle size, shear strength, moisture/density relationship, permeability and compressibility.

Rocks, soils, made-up ground and fill

Ground is the term used for the earth's surface which varies in the composition within the following five groups:

- Rocks
- Non-cohesive soils
- Cohesive soils
- Peat and organic soils
- Made-up ground and fill

Rocks include the hard, rigid, strongly cemented geological deposits such as granite, sandstone and limestone, and soils include the comparatively soft, loose, uncemented geological deposits such as gravel, sand and clay. Unlike rocks, soils and made up ground and fill are compacted under the compression of the loads of buildings on foundations.

Rocks

Rocks may be classified as sedimentary, metamorphic and igneous according to their geological formation as shown in Table 2, Volume 1, or in the five groups set out in Table 1, Volume 1, by reference to their presumed bearing value, which is the net loading intensity considered appropriate to the particular type of ground for preliminary design purposes. The presumed bearing values are based on the assumption that foundations are carried down to unweathered rock.

Hard igneous and gneissic rocks in sound condition have so high an allowable bearing pressure that there is little likelihood of foundation failure.

Hard limestones and hard sandstones are, when massively bedded, stronger than good quality concrete and it is rare that their full bearing capacity is utilised. Limestones are liable to solution by ground water containing dissolved carbon dioxide flowing along joints in the stone which may become enlarged and reduce the soundness of the rock as a foundation.

Schists and slates are rocks with pronounced cleavage. If the beds are shattered or steeply inclined a reduction in bearing values is made.

Hard shales and hard mudstones, formed from clayey or silty deposits by intense natural compaction, have a fairly high allowable bearing pressure.

Soft sandstones have a very variable allowable bearing pressure depending on the cementing material.

Soft shales and soft mudstones are intermediate between hard cohesive soils and rocks. They are liable to swell on exposure to water and soften.

Chalk and soft limestone include a variety of materials composed mainly of calcium carbonate and the allowable bearing pressure may vary widely. When exposed to water or frost these rocks deteriorate and should, therefore, be protected with a layer of concrete as soon as the final excavation level is reached.

Thinly bedded limestones and sandstones which are stratified rocks, often separated by clays or soft shales, have a variable allowable bearing pressure depending on the nature of the separating material.

Heavily shattered rocks have been cracked and broken up by natural processes. The allowable bearing pressure is determined by examination of loading tests.

Soils

The characteristics of a soil that affect its behaviour as a foundation are compressibility, cohesion of particles, internal friction and permeability. It is convenient to compare the characteristics and behaviour of clean sand, which is a coarse grained non-cohesive soil, with clay which is a fine grained cohesive soil, as foundations to buildings.

Compressibility

Under load sand is only slightly compressed due to the expulsion of water and some rearrangement of the particles. Because of its high permeability sand is rapidly compressed due to quick expulsion of water, and compression of sand subsoils keeps pace with the erection of buildings so that once the building is completed no further compression takes place.

Clay is very compressible, but due to its impermeability compression takes place slowly because of the very gradual expulsion of water through the narrow capillary channels in the clay. The compression of a clay subsoil under the foundation of a building may continue for some years after the building is completed, with consequent gradual settlement.

Cohesion of particles (plasticity)

There is negligible cohesion between the particles of sand and in consequence it is not plastic. There is marked cohesion between the particles of clay, which is plastic and can be moulded, particularly when wet. The different properties of compressibility and plasticity of sand and clay are commonly illustrated

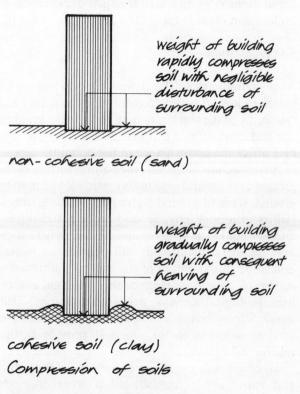

weight of building rapidly compresses soil with negligible disturbance of surrounding soil

non-cohesive soil (sand)

weight of building gradually compresses soil with consequent heaving of surrounding soil

cohesive soil (clay)
Compression of soils

Fig. 1

when walking over these soils. A foot makes a quick indent in sand with little disturbance of the soil around the imprint, whereas a foot sinks gradually into clay with appreciable heaving of the soil around the imprint. In like manner, the weight of a building on sand causes rapid compression with little disturbance of the surrounding soil, whereas on clay compression is slow and often accompanied by surrounding surface heave as illustrated in Fig. 1. The surface heave may be pronounced if the shear resistance of the clay is overcome as explained later.

Internal friction

There is considerable friction between the coarse particles of sand which strongly resists displacement or rearrangement of the particles. When this internal friction is overcome, for example by too great a load from the foundations of a building, the soil shears and suddenly gives way.

There is little friction between the fine particles of clay. Owing to the plastic nature of clay, shear failure, under the load of a building may take place along several strata simultaneously with consequent heaving of the soil as illustrated in Fig. 2. The shaded wedge of soil below the building is pressed down and displaces soil at both sides which moves along the slip surfaces indicated. In practice the load on a foundation may not be uniform over its area and the internal friction of the subsoil under the building may vary so that shear of the soil may occur on one side only as

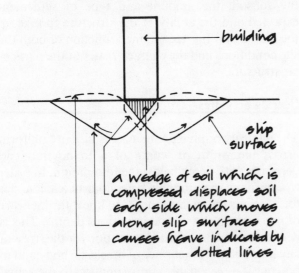

a wedge of soil which is compressed displaces soil each side which moves along slip surfaces & causes heave indicated by dotted lines

Plastic failure of soil

Fig. 2

illustrated in Fig. 3. This is an extreme, theoretical type of failure of a clay subsoil which is commonly used by engineers to calculate the resistance to shear of clay subsoils and presumes that the half cylinder of soil ABC rotates about centre O on slip plane ABC.

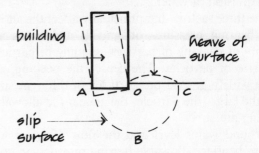

due to weight of building a cylinder of soil rotates causing heave of surface at one side

Plastic failure of soil

Fig. 3

Permeability

When water can pass rapidly through the pores or voids of a soil, the soil is said to be permeable. Coarse grained soils such as gravel and sand are permeable, and because water can drain rapidly through them they consolidate rapidly under load.

Fine grained soils such as clay have low permeability and because water passes very slowly through the pores they consolidate slowly.

Frost heave

The expansion of water in soils due to freezing atmospheric conditions was described in Volume 1. The expansion and consequent heaving of the soil occurs at the surface and for a depth of some 600 particularly in silty soils. The foundations of large buildings are generally some metres below the surface, at which level frost heave will have no effect in this country.

Non-cohesive coarse grained soils

Non-cohesive coarse grained soils such as gravels and sands consist of coarse grained, largely siliceous unaltered products of rock weathering. Gravels and sands composed of hard mineral particles have no plasticity and tend to lack cohesion especially when dry. Under pressure from the loads on foundations the soils in this group compress and consolidate rapidly by some rearrangement of the particles and the expulsion of water.

The three factors that principally affect the allowable bearing pressures on gravels and sands are density of packing of particles, grading of particles and size of particles. The denser the packing, the more widely graded the particles of different sizes and the larger the particles the greater the allowable bearing pressure.

Ground water level and the flow of water will adversely affect allowable bearing pressures in non-cohesive soils where ground water level is near to the foundation level and so affect the density of packing and where flow of water may wash out finer particles and so affect grading.

Non-cohesive soils must be laterally confined to prevent spread of the soil under pressure.

Cohesive fine grained soils

Cohesive fine grained soils such as clays and silts are a natural deposit of the finer siliceous and aluminous products of rock weathering. Clay is smooth and greasy to the touch, shows high plasticity, dries slowly and shrinks appreciably on drying. The principal characteristic of cohesive soils as a foundation is their susceptibility to slow volume changes. Under the pressure of the load on foundations, clay soils are very gradually compressed by the expulsion of water or water and air through the very many fine capillary paths so that buildings settle gradually during erection and this settlement will continue for some years after the building is completed.

Seasonal variations in ground water and vigorous growth of trees and shrubs will cause appreciable shrinkage, drying and wetting expansion of cohesive soils. Shrinkage and expansion due to seasonal variations will extend to one metre or more in periods of severe drought below the surface in Great Britain and up to four metres or more below large trees. When shrubs and trees are removed to clear a site for building on cohesive soils, for some years after the clearance there will be ground recovery as the soil gradually recovers water previously taken out by trees and shrubs. This gradual recovery of water by cohesive soils and consequent expansion may take several years.

Volume changes in cohesive soils under or around foundations close to the external faces of exposed buildings will be greater due to seasonal variations than under buildings which give protection from the effects of seasonal variations.

Peat and organic soils

Peat and organic soils have a high proportion of fibrous or spongy vegetable matter from the decay of plants mixed with varying proportions of fine sand, silt or clay. These soils are highly compressible and will not serve as a stable foundation for buildings.

Made-up ground and fill

Made-up ground and fill will not usually serve as a stable foundation for buildings due to the extreme variability of the materials used to make up ground and the variability of the compaction or natural settlement of these materials.

Foundation design

Failure of the foundation of a building may be due to excessive settlement by compaction of subsoil, collapse of subsoil by failure in shear or differential settlement of different parts of the foundation. The allowable bearing pressure intensity at the base of foundations is the maximum allowable net loading taking into account the ultimate bearing capacity of the subsoil, the amount and type of settlement expected and the ability of the structure to take up the settlement. It is a combined function of both the site conditions and the characteristics of the particular structure.

Bearing pressures

The intensity of pressure on a subsoil is not uniform across the width or length of a foundation and decreases with depth below the foundation. In order to determine the probable behaviour of a soil under foundations the engineer needs to know the intensity of pressure on the subsoil at various depths. This is determined by Boussinesq's equation for the stress at any point below the surface of an elastic body and in practice is a reasonable approximation to the actual stress in soil. By applying the equation, the vertical stress on planes at various depths below a point can be calculated and plotted as shown in Fig. 4. The

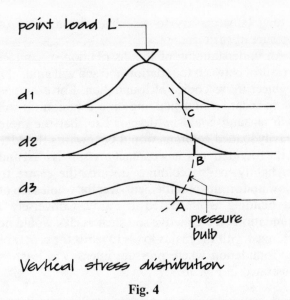

Vertical stress distribution

Fig. 4

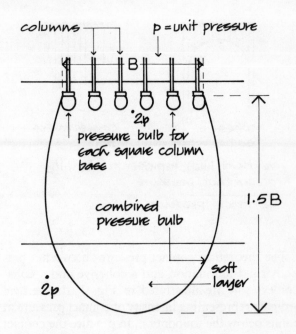

Combined pressure bulb

Fig. 6

vertical ordinates at each level d_1, d_2, etc., represent graphically unit stress at points at that level. If points of equal stress, A, B, and C are joined the result is a bulb of unit pressure extending down from L. If this operation is repeated for unit area under a foundation the result is a series of bulbs of equal unit pressure as illustrated in Fig. 5.

Thus the bulb of pressure gives an indication of likely stress in subsoils at various points below a foundation. If there are separate foundations close together, as for example where there is a group of columns, then the bulbs of pressure can be combined to form one large pressure bulb diagram as illus-

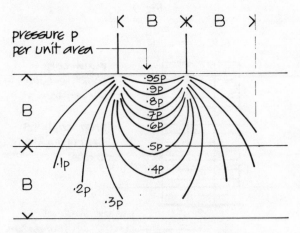

Bulbs of vertical pressure under a strip foundation

Fig. 5

trated in Fig. 6. Where bulbs of pressure of adjacent foundations intersect an increased intensity of pressure occurs.

It is because there are often strata of different soils below the surface that some knowledge of the intensity of pressure is necessary. For example, the soft layer in Fig. 6 is intersected by the combined pressure bulb; that may indicate that unit pressure is so great that the soft layer may fail. In practice it would be tedious to construct a bulb of pressure diagram each time a foundation were to be designed and engineers today generally employ ready prepared diagrams or charts to determine pressure intensities below foundations.

Contact pressure

A perfectly flexible foundation uniformly loaded will cause uniform contact pressure with all types of soil. A perfectly flexible foundation supposes a perfectly flexible structure supporting flexible floors, roof and cladding. The CLASP system of building commonly employed for schools uses a flexible frame (see Chapter 2) and was originally designed to accommodate movement in the foundation of buildings on land subject to mining subsidence. Most large buildings, however, have rigid foundations designed to support a rigid or semi-rigid frame.

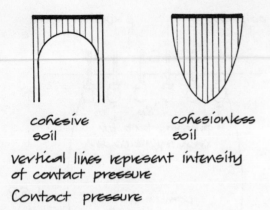

cohesive soil   cohesionless soil

vertical lines represent intensity of contact pressure

Contact pressure

Fig. 7

The theoretical contact pressures between a perfectly rigid foundation and a cohesive and a cohesionless soil are illustrated in Fig. 7, the vertical ordinate representing intensity of contact pressure at points below the foundation. In practice the contact pressure on a cohesive soil such as clay is reduced at the edges of the foundation by yielding of the clay and as the load on the foundation increases more yielding of the clay takes place so that the stresses at the edges decrease and those at the centre of the foundation increase as illustrated in Fig. 8.

The contact pressure on a cohesionless soil such as dry sand remains parabolic as illustrated in Fig. 8 and the maximum intensity of pressure increases with increased load. If the foundation is below ground the edges stresses are no longer zero as illustrated in Fig. 8 and increase with increase of depth below ground.

For footings the assumption is made that contact pressure is distributed uniformly over the effective area of the foundation as differences in contact pressure are usually covered by the margin of safety used in design.

For large spread foundations and raft foundations

it may be necessary to calculate the intensity of pressure at various depths.

An understanding of the distribution of contact pressures between foundation and soil will guide the engineer in his choice of foundation. For example, the foundation of a building on a cohesionless soil such as sand could be designed so that the more heavily loaded columns would be towards the edge of the foundation where contact pressure is least and the lightly loaded columns towards the centre to allow uniformity of settlement over the whole area of the building as illustrated in Fig. 9. Conversely a foundation on a cohesive soil such as clay would be arranged with the major loads towards the centre of the foundation where pressure intensity is least as illustrated in Fig. 10.

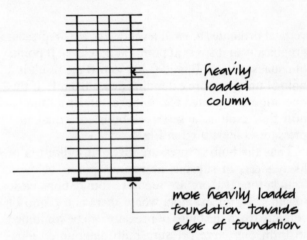

heavily loaded column

more heavily loaded foundation towards edge of foundation

Foundation on a cohesionless subsoil

Fig. 9

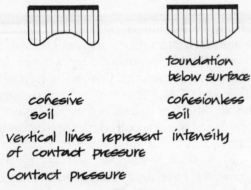

foundation below surface

cohesive soil   cohesionless soil

vertical lines represent intensity of contact pressure

Contact pressure

Fig. 8

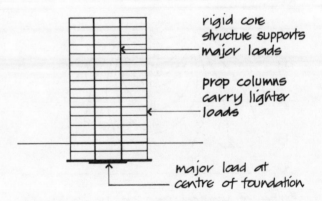

rigid core structure supports major loads

prop columns carry lighter loads

major load at centre of foundation

Foundation on a cohesive subsoil

Fig. 10

Differential settlement (relative settlement)

Parts of the foundation of a building may suffer different magnitudes of settlement due to variations in load on the foundations or variations in the subsoil and different rates of settlement due to variations in the subsoil. These variations may cause distortion of a rigid or semi-rigid frame and consequent damage to rigid infill panels and cracking of loadbearing walls, rigid floors and finishes. Some degree of differential settlement is inevitable in the foundation of most buildings but so long as this is not pronounced or can be accommodated in the design of the building the performance of the building will not suffer.

The degree to which differential settlement will adversely affect a building depends on the structural system employed. Solid load bearing brick and masonry walls can accommodate small differential settlement through small hair cracks opening in mortar joints between the small units of brick or stone. These cracks, which are not visible, do not weaken the structure nor encourage the penetration of rain. More pronounced differential settlement such as is common between the main walls of a house and the less heavily loaded bay window bonded to it, may cause visible cracks in the brickwork at the junction of the bay window and the wall. Such cracks will allow rain to penetrate the thickness of the wall. To avoid this, either the foundation should be strengthened or some form of slip joint be formed at the junction of the bay and the main wall.

High, framed buildings are generally designed as rigid or semi-rigid structures and any appreciable differential settlement should be avoided. Differential settlement of more than 25 between adjacent columns of a rigid or semi-rigid framed structure may cause such serious racking of the frame that local stress at the junction of vertical and horizontal members of the frame may endanger the stability of the structure and also crack solid panels within the frame. An empirical rule employed by engineers in the design of foundations is to limit differential settlement between adjacent columns to $\frac{1}{500}$th of the distance between them.

Differential settlement can be reduced by a stiff structure or substructure or a combination of both. A deep hollow box raft (see later in Fig. 17) has the advantage of reducing net loading intensity and producing more uniform settlement.

A common settlement problem occurs in modern buildings where a tower or slab block is linked to a low podium. Plainly there will tend to be a more pronounced settlement of the foundations of the tower or slab block than the podium and at the junction of the two structures there must be structural discontinuity and some form of flexible joint that will accommodate the differences in settlement. Figure 11 illustrates two examples of this arrangement.

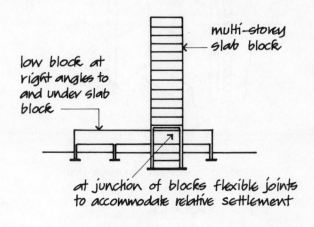

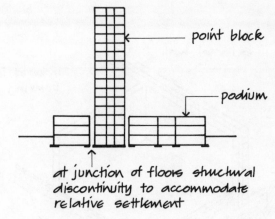

Relative settlement

Fig. 11

FOUNDATIONS

Foundations may be classified as:

- Strip foundations
- Pad foundations
- Raft foundations
- Pile foundations

and these are illustrated in Fig. 12.

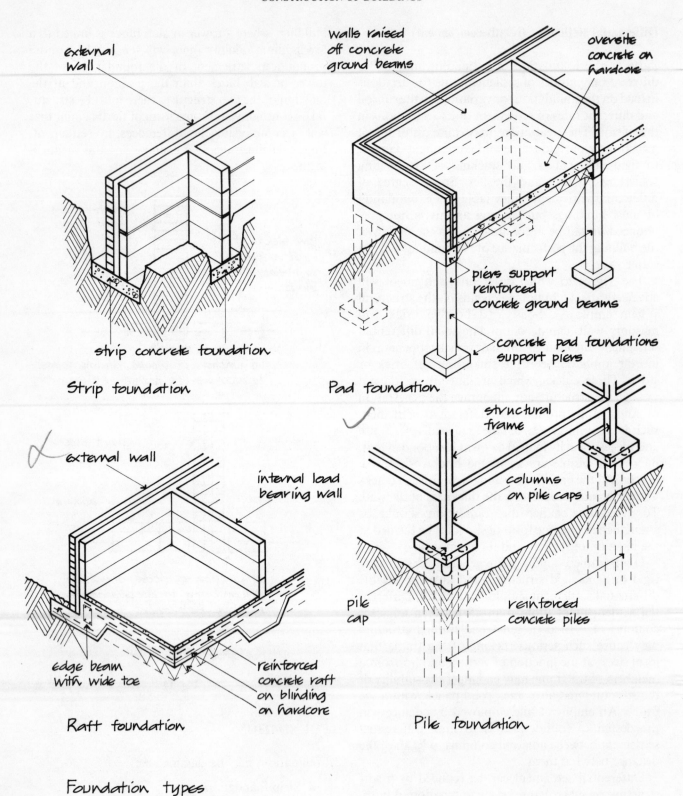

external wall

strip concrete foundation

Strip foundation

walls raised off concrete ground beams

oversite concrete on hardcore

piers support reinforced concrete ground beams

concrete pad foundations support piers

Pad foundation

external wall

internal load bearing wall

edge beam with wide toe

reinforced concrete raft on blinding on hardcore

Raft foundation

structural frame

columns on pile caps

pile cap

reinforced concrete piles

Pile foundation

Foundation types

Fig. 12

10

Strip foundations

Strip foundations (see Volume 1, Chapter 1) consist of a continuous, longitudinal strip of concrete designed to spread the load from uniformly loaded walls of brick, masonry or concrete to a sufficient area of subsoil. The spread of the strip depends on foundation loads and the bearing capacity and shear strength of the subsoil. The thickness of the foundation depends on the strength of the foundation material. Strip foundations with a wide spread are commonly of reinforced concrete, as illustrated in Fig. 13.

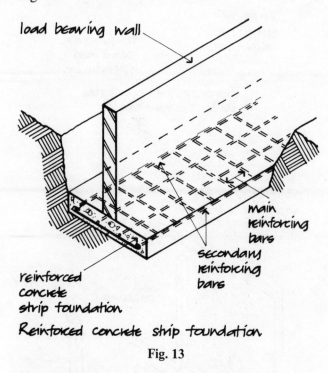

Reinforced concrete strip foundation

Fig. 13

Pad foundations

The foundation to piers of brick, masonry and reinforced concrete and steel columns is often in the form of a square or rectangular isolated pad of concrete to spread a concentrated load. The area of this type of foundation depends on the load on the foundation and the bearing and shear strength of the subsoil and its thickness on the strength of the foundation material. The simplest form of pad foundation consists of a pad of mass concrete as illustrated in Volume 1, illustrating a pier and foundation beam base for a small building. Heavily loaded pad foundations supporting columns of framed buildings are generally of reinforced concrete

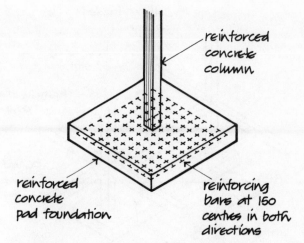

Reinforced concrete pad foundation

Fig. 14

as illustrated in Fig. 14 showing the base of a reinforced concrete column. The area of the pad foundation is determined by the load of the foundation and the allowable bearing pressure on the subsoil and the thickness and reinforcement from a calculation of bending and shear stresses.

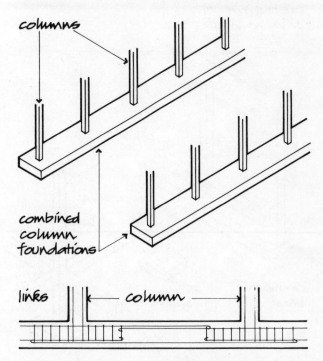

longitudinal section of foundation

Combined column foundation

Fig. 15

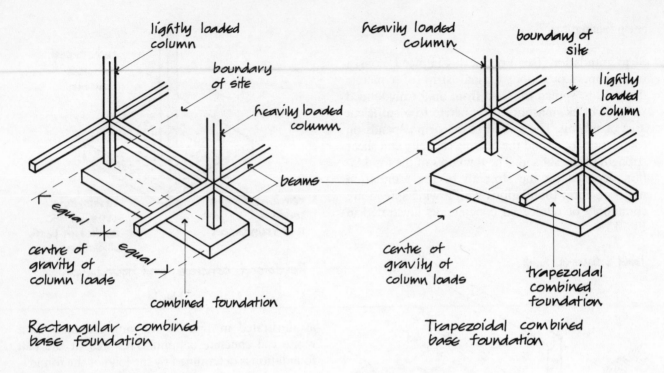

lightly loaded column

boundary of site

heavily loaded column

beams

equal   equal

centre of gravity of column loads

combined foundation

Rectangular combined base foundation

heavily loaded column

boundary of site

lightly loaded column

beams

centre of gravity of column loads

trapezoidal combined foundation

Trapezoidal combined base foundation

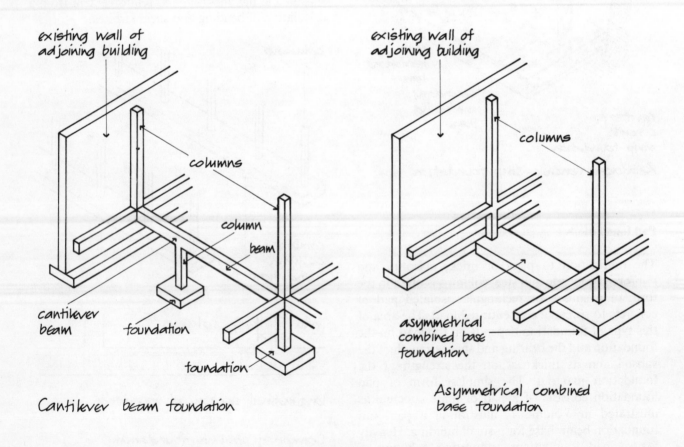

existing wall of adjoining building

columns

column

beam

cantilever beam

foundation

foundation

Cantilever beam foundation

existing wall of adjoining building

columns

asymmetrical combined base foundation

Asymmetrical combined base foundation

Fig. 16

12

Where there is a wide spread of pad foundations to a framed building due to the low bearing capacity of the subsoil or the close spacing of columns, such that the edge of adjacent separate foundations would be close together, it may be economical and convenient to form one continuous column foundation, as illustrated in Fig. 15. This in effect is a reinforced concrete strip foundation supporting concentrated loads.

Combined foundations

The foundations of adjacent columns are combined

(1) when a column is so close to the boundary of the site that a separate foundation would be eccentrically loaded and (2) where foundations of adjacent columns are linked to resist uplift, overturning or opposing forces.

Because the base of the column adjacent to the site boundary cannot spread uniformly around the column, it is combined with the base of an adjacent column to form a combined or balanced base foundation as illustrated in Fig. 16.

Where a framed building is to be erected alongside an existing building it is often necessary to use a cantilever or asymmetrical combined base founda-

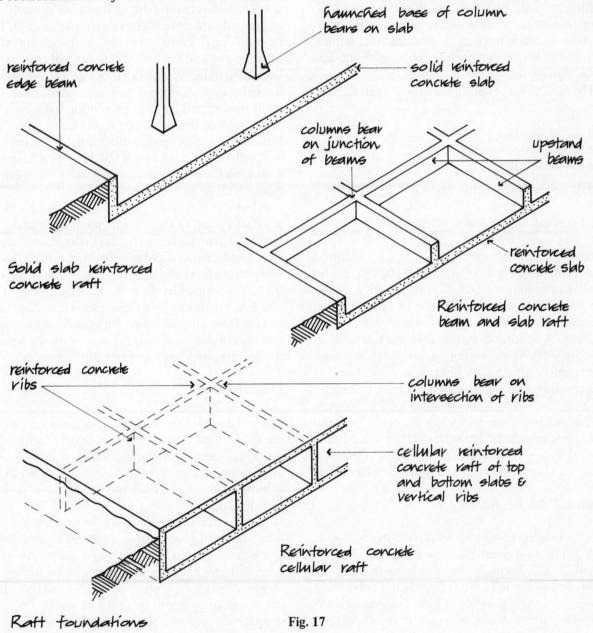

Raft foundations   **Fig. 17**

13

tion for columns next to the existing building, so that pressure on the subsoil due to the base may not so heavily surcharge the subsoil under the foundation of the existing building as to cause it to settle appreciably. Cantilever and asymmetrical combined foundations are illustrated in Fig. 16.

Raft foundations

A raft foundation is continuous in two directions, usually covering an area equal to or greater than the base of a building or structure. Raft foundations are used for lightly loaded structures on soils with poor bearing capacity or where variations in soil conditions necessitate a considerable spread of the load, for heavier loads in place of isolated foundations, where differential settlements are significant and where mining subsidence is likely.

The three types of reinforced concrete raft foundations are:

● Solid slab raft
● Beam and slab raft
● Cellular raft

and these are illustrated in Fig. 17.

Solid slab raft foundation

Solid slab raft foundation is a solid reinforced concrete slab generally of uniform thickness, cast on subsoils of poor or variable bearing capacity, so that the loads from walls or columns of lightly loaded structures are spread over the whole area of the building. A solid slab raft of uniform thickness to support walls and a variant, a solid slab raft with stiffening edge beams are illustrated in Volume 1.

The solid slab raft foundation illustrated in Fig. 17 supports reinforced concrete columns. The columns have haunched bases to spread the point load and resist punching shear. The solid slab raft is cast below ground level with an upstand edge beam. The lower floor will take the form of a suspended timber floor.

Beam and slab raft foundation

As a foundation to support the heavier loads of walls or columns a solid slab raft would require considerable thickness. To make the most economical use of reinforced concrete in a raft foundation supporting heavier loads it is practice to form a beam and slab raft. This raft consists of upstand or downstand beams that take the loads of walls or columns and spread them to the monolithically cast slab which bears on natural subsoil.

On compact soils which can be excavated without the necessity of timbering to trenches it is economical to use downstand beams, illustrated in Fig. 18, and where the subsoil is granular, upstand beams may be necessary, also illustrated in Fig. 18.

Cellular raft foundation

Where differential settlements are likely to be significant and the foundations have to support considerable loads the great rigidity of the monolithically cast reinforced concrete cellular raft is an advantage. This type of raft consists of top and bottom slabs separated by and reinforced with vertical cross ribs in both directions, as illustrated in Fig. 17. The monolithically cast reinforced concrete cellular raft has great rigidity and spreads foundation loads over the whole area of the substructure to reduce consolidation settlement and avoid differential settlement.

A cellular raft may be the full depth of a basement storey and the cells of the raft may be used for mechanical plant and storage.

A cellular raft is also used when deep basements are constructed to reduce settlement by utilising the overburden pressure that occurs in deep excavations. This negative or upward pressure occurs in the bed of deep excavations in the form of an upward heave of the subsoil caused by the removal of the overburden, which is taken out by excavation. This often quite considerable upward heave can be utilised to counteract consolidation settlement caused by the load of the building and so reduce overall settlement.

Pile foundations

The word pile is used to describe columns, usually of reinforced concrete, driven or cast in the ground in order to carry foundation loads to some deep underlying firm stratum or to transmit loads to the subsoil by the friction of their surfaces in contact with the subsoil.

The main function of a pile is to transmit loads to lower levels of ground by a combination of friction along their sides and end bearing at the pile point or base. Piles that transfer loads mainly by friction to clays and silts are termed *friction piles* and those that mainly transfer loads by end bearing to compact gravel, hard clay or rock are termed *end-bearing piles.*

14

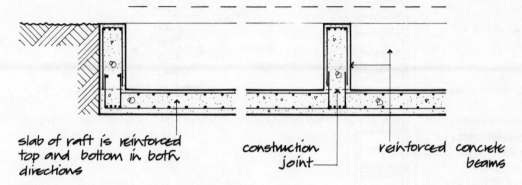

the floor is constructed with
precast reinforced concrete beams
bearing on upstand beams of raft

slab of raft is reinforced
top and bottom in both
directions

construction
joint

reinforced concrete
beams

Beam and slab raft with upstand beams

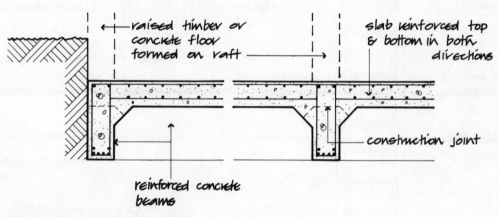

raised timber or
concrete floor
formed on raft

slab reinforced top
& bottom in both
directions

reinforced concrete
beams

construction joint

Beam and slab raft with downstand beams

Reinforced concrete beam and slab rafts

Fig. 18

Piles may be classified by their effect on the subsoil as *displacement piles* or *non-displacement piles*. *Displacement piles* are driven or otherwise forced into the ground to displace subsoil, such as solid piles and piles formed inside tubes which are driven into the ground and which are closed at their lower end by a shoe or plug which may either be left in place or extruded to form an enlarged toe. *Non-displacement piles* are formed by boring or other methods of excavation that do not substantially displace subsoil and where the borehole is lined with a casing or tube that is either left in place or extracted as the hole is filled.

Driven piles are those formed by driving a precast pile and those made by casting concrete in a hole formed by driving. *Bored piles* are those formed by casting concrete in a hole previously bored or drilled in the subsoil.

Driven piles

Square, polygonal or round section reinforced concrete piles are cast in moulds in the manufacturer's yard and are cured to develop maximum strength. The placing of the reinforcement and the mixing, placing compaction and curing of the concrete can be

15

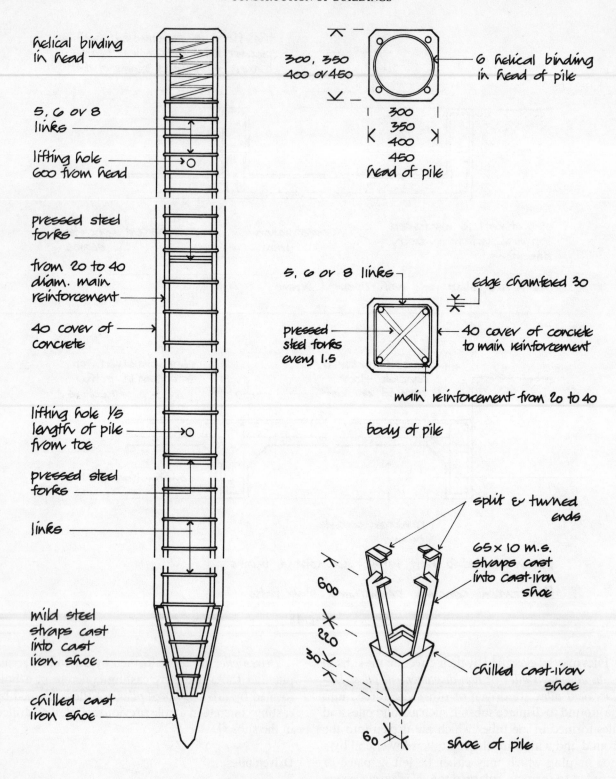

helical binding in head

5, 6 or 8 links

lifting hole 600 from head

pressed steel forks

from 20 to 40 diam. main reinforcement

40 cover of concrete

lifting hole ⅓ length of pile from toe

pressed steel forks

links

mild steel straps cast into cast iron shoe

chilled cast iron shoe

300, 350 400 or 450

6 helical binding in head of pile

300
350
400
450
head of pile

5, 6 or 8 links

edge chamfered 30

pressed steel forks every 1.5

40 cover of concrete to main reinforcement

main reinforcement from 20 to 40

body of pile

split & turned ends

65 × 10 m.s. straps cast into cast-iron shoe

600

220

30

chilled cast-iron shoe

60

shoe of pile

Precast reinforced concrete piles

Fig. 19

16

accurately controlled to produce piles of uniform strength and cross-section. The piles are lifted into position and driven into the ground by means of a mechanically operated drop hammer attached to a mobile piling rig. Figure 19 is an illustration of a typical pile.

The pile is driven in until a predetermined 'set' is reached. The word set is used to describe the distance that a pile is driven into the ground by the force of the hammer falling a measurable distance. From the weight of the hammer and the distance it falls the resistance of the ground can be calculated and the bearing capacity of the pile calculated.

To connect the top of the precast pile to the reinforced concrete foundation the top 300 of the length of the pile is broken to expose reinforcement to which the reinforcement of the foundation is connected. Precast driven piles are not in general used on sites in built-up areas due firstly to difficulties in moving them through narrow streets and secondly to the nuisance caused by the noise of driving and the vibration caused by driving which might damage adjacent buildings. Driven piles are used as end-bearing piles in weak subsoils where they are driven to a firm underlying stratum. Driven piles give little strength in bearing due to friction of their sides in contact with soil, particularly when the surrounding soil is clay. This is due to the fact that the operation of driving moulds the clay around the pile and so reduces frictional resistance between the pile and the surrounding clay.

In coarse grained cohesionless soils where the piles do not reach a firm stratum, driven piles act as friction bearing piles due to the action of pile driving, which compacts the coarse particles around the sides of the pile and so increases frictional resistance and in compacting the soil increases its strength. This type of piled foundation is sometimes described as a floating foundation, as is a cast-in-place piled foundation, as bearing is mainly by friction and in effect the piles are floating in the subsoil rather than bearing on firm soil.

Prestressed concrete piles may be used in place of cast concrete piles. The advantages of a prestressed concrete pile are that, due to the prestress of the concrete the pile will have a smaller cross-sectional area than a comparable cast pile and the prestress will reduce tensile cracking of the pile and so give greater durability, particularly in water bearing soils.

Timber piles are little used for the foundations of buildings because of the difficulty of obtaining sufficiently large sections and lengths and because of the possibility of timber rotting underground.

Driven cast-in-place piles

Driven cast-in-place piles are of two types, the first has a permanent steel or concrete casing and the second is without permanent casing. The purpose of driving and maintaining a permanent casing is to consolidate the subsoil around the pile casing by the action of driving, and the lining is left in place to protect the concrete cast inside the lining against weak strata of subsoil that might otherwise fall into the pile excavation and to protect the green concrete of the pile against static or running water.

Figure 20 is an illustration of a driven cast-in-place pile with a permanent reinforced concrete casing. Precast reinforced concrete shells are threaded on a steel mandrel. Metal bands and bitumen seal joints between shells. The mandrel and shells are lifted on to the piling rig and then driving into the ground. At the required depth the mandrel is removed, a reinforcing cage is lowered into the shells and the pile completed by casting concrete inside the shells. This type of pile is used principally in soils of poor bearing capacity and saturated soils where the concrete shells protect the green concrete cast inside them, from static or running water.

A driven cast-in-place pile without permanent casing is illustrated in Fig. 21. The base of a steel lining tube, supported on a piling rig, is filled with ballast. A drop hammer rams the ballast and the tube into the ground and at the required depth the tube is restrained and the ballast is hammered in to form an enlarged toe as shown. Concrete is placed by hammering it inside the lining tube which is gradually withdrawn. The effect of driving the tube and the ballast into the ground is to compact the soil around the pile and the subsequent hammering of the concrete consolidates it into pockets and weak strata. The enlarged toe provides additional bearing area at the base of the pile. This type of pile acts mainly as a friction pile.

Another type of driven cast-in-place pile without permanent casing is formed by driving a lining tube with cast iron shoe into the ground with a piling hammer operating in a piling rig as illustrated in Fig. 22. Concrete is placed by hammering the lining tube as it is withdrawn. The particular application of this type of pile is for piles formed through a substratum so compact as to be incapable of being taken out by

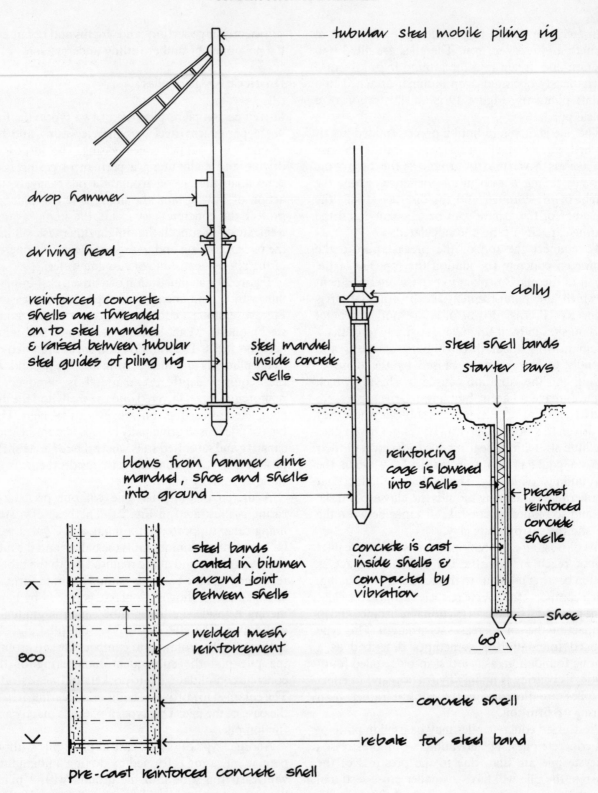

tubular steel mobile piling rig

drop hammer

driving head

reinforced concrete shells are threaded on to steel mandrel & raised between tubular steel guides of piling rig

steel mandrel inside concrete shells

blows from hammer drive mandrel, shoe and shells into ground

dolly

steel shell bands

starter bars

reinforcing cage is lowered into shells

precast reinforced concrete shells

concrete is cast inside shells & compacted by vibration

shoe

60°

steel bands coated in bitumen around joint between shells

welded mesh reinforcement

800

concrete shell

rebate for steel band

pre-cast reinforced concrete shell

Driven cast-in-place pile

Fig. 20

18

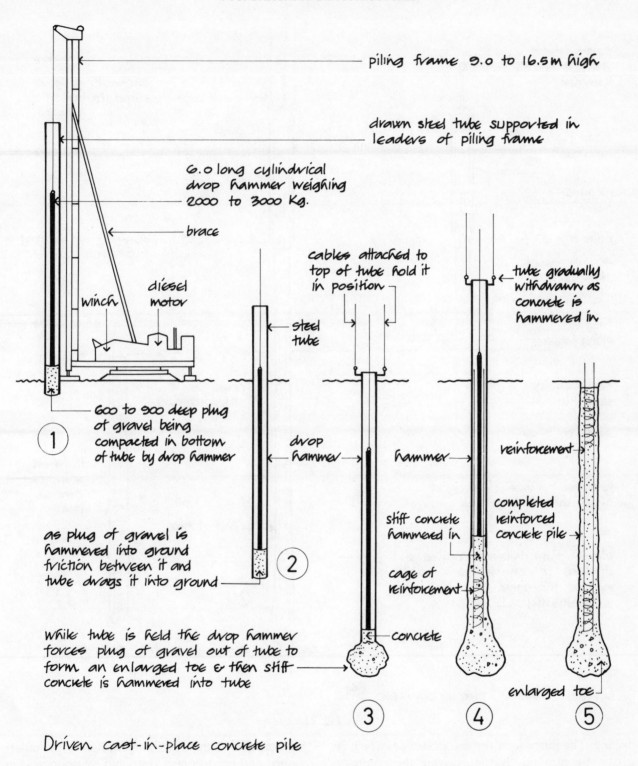

piling frame 9.0 to 16.5 m high

drawn steel tube supported in leaders of piling frame

6.0 long cylindrical drop hammer weighing 2000 to 3000 Kg.

brace

winch

diesel motor

cables attached to top of tube hold it in position

steel tube

drop hammer

hammer

tube gradually withdrawn as concrete is hammered in

reinforcement

completed reinforced concrete pile

stiff concrete hammered in

cage of reinforcement

concrete

enlarged toe

① 600 to 900 deep plug of gravel being compacted in bottom of tube by drop hammer

as plug of gravel is hammered into ground friction between it and tube drags it into ground

②

while tube is held the drop hammer forces plug of gravel out of tube to form an enlarged toe & then stiff concrete is hammered into tube

③ ④ ⑤

Driven cast-in-place concrete pile

Fig. 21

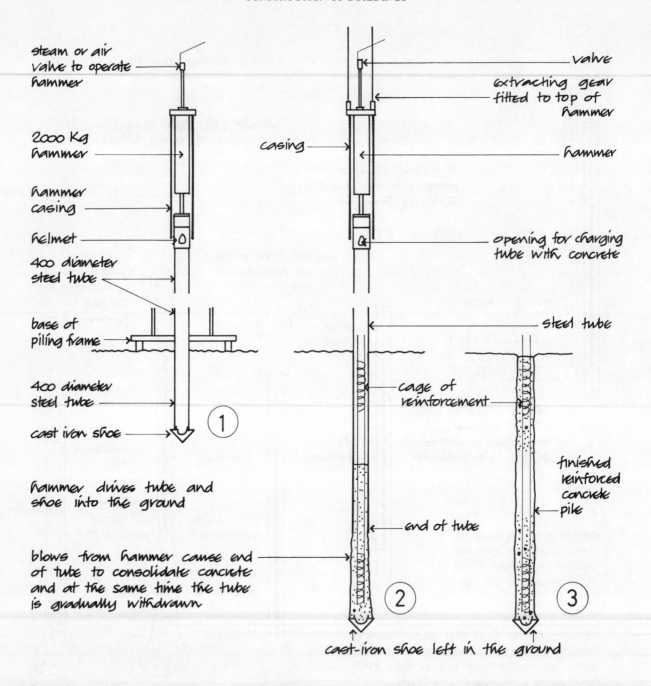

steam or air valve to operate hammer

2000 Kg hammer

hammer casing

helmet

400 diameter steel tube

base of piling frame

400 diameter steel tube

cast iron shoe

hammer drives tube and shoe into the ground

blows from hammer cause end of tube to consolidate concrete and at the same time the tube is gradually withdrawn

valve

extracting gear fitted to top of hammer

casing

hammer

opening for charging tube with concrete

steel tube

cage of reinforcement

finished reinforced concrete pile

end of tube

cast-iron shoe left in the ground

Driven cast-in-place concrete pile

Fig. 22

drilling. The purpose of the cast iron shoe, which is left in the ground, is to penetrate the compact stratum through which the pile is formed.

Jacked piles

Figure 23 illustrates a system of jacked piles that are designed for use in cramped working conditions, as

for example where an existing wall is to be underpinned and headroom is restricted by floors and in situations where the vibration caused by pile driving might damage existing buildings.

Where the wall to be underpinned has a sound concrete base the pile sections are jacked into the ground under the base, as illustrated in Fig. 23, and a concrete cap is cast on top of the pile and up to the

20

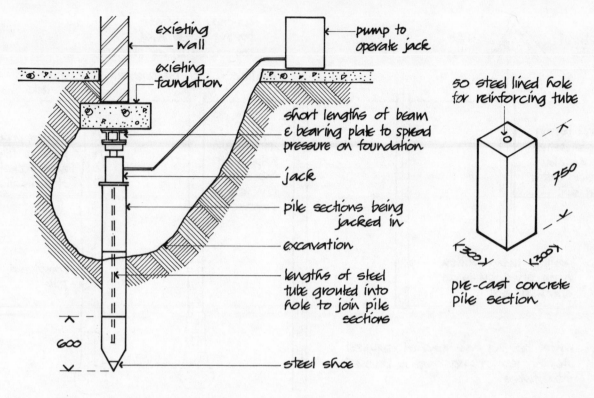

Pre-cast concrete jacked pile for underpinning

Fig. 23

underside of the concrete base. When the wall to be underpinned has a poor base and the wall might be disturbed by jacking piles under it, then pairs of piles are jacked in each side of the wall to support steel or reinforced concrete needles that in turn support the weight of the wall.

The precast concrete sections are jacked into the ground, as illustrated, and lengths of tube are grouted inside the steel lined hole in each section to make a strong connection between sections.

Piles formed on both sides of the wall are jacked in against units loaded with kentledge.

Bored piles

A hole is bored or drilled by means of earth drills or mechanically operated augers which withdraw soil from the hole into which the pile is to be cast. Usually steel lining tubes are lowered or knocked in, as the soil is taken out, to maintain the sides of the drilling. As the pile is cast the lining tubes are gradually withdrawn.

The principal advantages of bored piles are that light, easily manipulated equipment may be used for the work and that a precise analysis of the subsoil strata is obtained from the soil withdrawn during drilling. Disadvantages are that it is not possible to check that the concrete is adequately compacted and that there is adequate cover of concrete to reinforcement.

Figure 24 illustrates the drilling and casting of a bored cast-in-place pile. Soil is withdrawn from inside the lining tubes with a cylindrical clay cutter that is dropped into the hole, bites into the cohesive soil, is withdrawn and the soil knocked out of it. Coarse grained soil is withdrawn by dropping a shell cutter (or bucket) into the hole. Soil, retained on the upward hinged flap, is emptied when the cutter is withdrawn. The operation of boring the hole is more rapid than might be supposed and a pile can be bored and cast in a matter of hours.

Concrete is cast under pressure through a steel helmet which is screwed to the top of the lining tubes. The application of air pressure at once compacts the concrete and simultaneously lifts the helmet and lining tubes as the concrete is compacted. As the

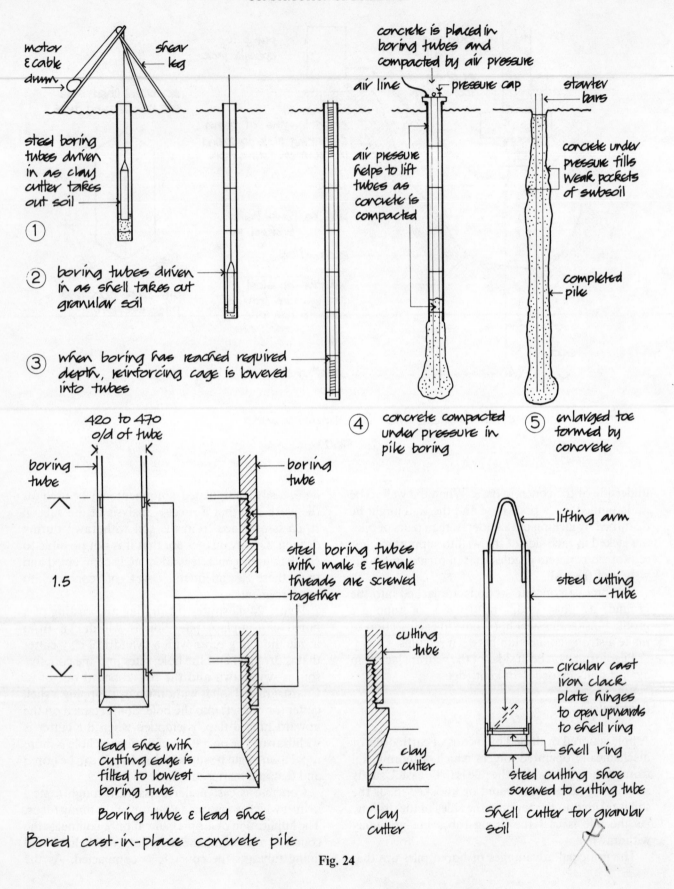

motor & cable drum

shear leg

steel boring tubes driven in as clay cutter takes out soil

①

② boring tubes driven in as shell takes out granular soil

③ when boring has reached required depth, reinforcing cage is lowered into tubes

concrete is placed in boring tubes and compacted by air pressure

air line

pressure cap

starter bars

air pressure helps to lift tubes as concrete is compacted

concrete under pressure fills weak pockets of subsoil

completed pile

④ concrete compacted under pressure in pile boring

⑤ enlarged toe formed by concrete

420 to 470 o/d of tube

boring tube

boring tube

1.5

steel boring tubes with male & female threads are screwed together

lead shoe with cutting edge is fitted to lowest boring tube

Boring tube & lead shoe

cutting tube

clay cutter

Clay cutter

lifting arm

steel cutting tube

circular cast iron clack plate hinges to open upwards to shell ring

shell ring

steel cutting shoe screwed to cutting tube

Shell cutter for granular soil

Bored cast-in-place concrete pile

Fig. 24

lining tubes are withdrawn, protruding sections are unscrewed and the helmet refixed until the pile is completed.

As the concrete is cast under pressure it extends beyond the circumference of the original drilling to fill and compact weak strata and pockets in the subsoil, as illustrated in Fig. 24. Because of the irregular shape of the surface of the finished pile it acts mainly as a friction pile to form what is sometimes called a floating foundation.

Figure 25 illustrates the formation and casting of a large diameter bored pile. A tracked crane supports hydraulic rams and a diesel engine which operates a Kelly bar and rotary bucket drill. The diesel engine rotates the Kelly bar and bucket in the bottom in which angled blades excavate and fill the bucket with soil. The hydraulic rams force the bucket into the ground. The filled bucket is raised and emptied and drilling proceeds. In non-cohesive soils the excavation is lined with steel lining tubes. To provide increased end bearing the drill can be belled out to twice the diameter of the pile.

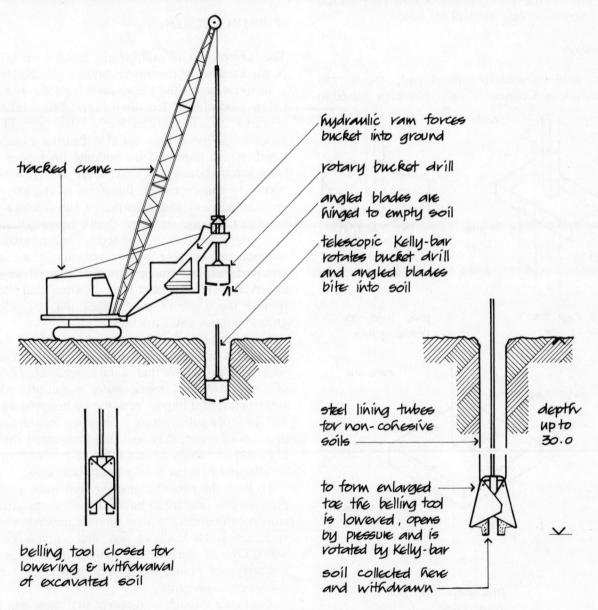

tracked crane

hydraulic ram forces bucket into ground

rotary bucket drill

angled blades are hinged to empty soil

telescopic Kelly-bar rotates bucket drill and angled blades bite into soil

steel lining tubes for non-cohesive soils

depth up to 30.0

belling tool closed for lowering & withdrawal of excavated soil

to form enlarged toe the belling tool is lowered, opens by pressure and is rotated by Kelly-bar

soil collected here and withdrawn

Large diameter bored cast-in-place reinforced concrete pile

Fig. 25

Spacing of piles

The spacing of piles should be wide enough to allow for the necessary number of piles to be driven or bored to the required depth of penetration without damage to adjacent construction or to other piles in the group. Piles are generally formed in comparatively close groups for economy in the size of the pile caps to which they are connected.

As a general rule the spacing, centre to centre of friction piles, should be not less than the perimeter of the pile and the spacing of end-bearing piles not less than twice the least width of the pile.

Pile caps

Piles may be used to support pad, strip or raft foundations. Commonly a group of piles is used to

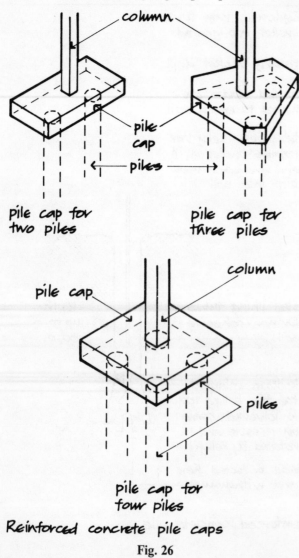

pile cap for two piles

pile cap for three piles

pile cap for four piles

Reinforced concrete pile caps

Fig. 26

support a column or pier base. The load from the column or pier is transmitted to the piles through a reinforced concrete pile cap which is cast over the piles. To provide structural continuity the reinforcement of the piles is linked to the reinforcement of the pile caps through starter bars protruding from the top of the cast-in-place piles or through reinforcement exposed by breaking off the top concrete from precast piles. Figure 26 illustrates typical arrangements of pile caps.

SUBSTRUCTURES

The substructure of multi-storey buildings is often constructed below natural or artificial ground level.

In towns and cities the ground for some metres below ground level has often been filled, over the centuries, to an artificial level. Filled ground is generally of poor and variable bearing capacity which is not improved by building operations to form a foundation at a lower level. It is generally necessary and expedient, therefore, to remove the artificial ground and construct a substructure or basement of one or more floors below ground. Similarly, where there is a top layer of natural ground of poor and variable bearing capacity, it is often removed and a substructure formed. Where there are appreciable differences of level on a building site a part or the whole of the building may be below ground level as a substructure.

The natural or artificial ground around the substructure is often permeable to water and may retain water to a level above that of the lower level or floor of a substructure. Ground water in soil around a substructure will impose pressure on both the walls and floor of a substructure. This, often considerable pressure of water, may well penetrate small shrinkage and movement cracks in dense concrete walls and floors and dense solidly built brick walls.

To limit the penetration of ground water under pressure it is practice to build in water stops across construction and movement joints in concrete walls and floors and to line brick walls and concrete floors with a layer of impermeable material in the form of a waterproof lining like a tank, hence the term 'tanking to basements'.

Another approach is to accept that there will be some penetration of ground water and construct an enclosing wall of dense concrete with water stops, separate from and around the substructure, so that water penetrating the outer walls drains to a sump outside the substructure enclosing walls.

Waterstops to concrete walls and floors

Dense concrete, which is practically impermeable to water, would by itself effectively exclude ground water were it possible to prevent shrinkage, constructional, structural, thermal and moisture movement cracks. As concrete dries out after placing it shrinks and this inevitable drying shrinkage causes cracks particularly at construction joints, through which ground water will penetrate. To minimise shrinkage cracks it is practice to cast adjacent bays or areas of concrete in floors and walls with a gap of 450 or 600 between them. When the bays of concrete have dried out for some days and much of the drying shrinkage has taken place, concrete is then cast into the spaces

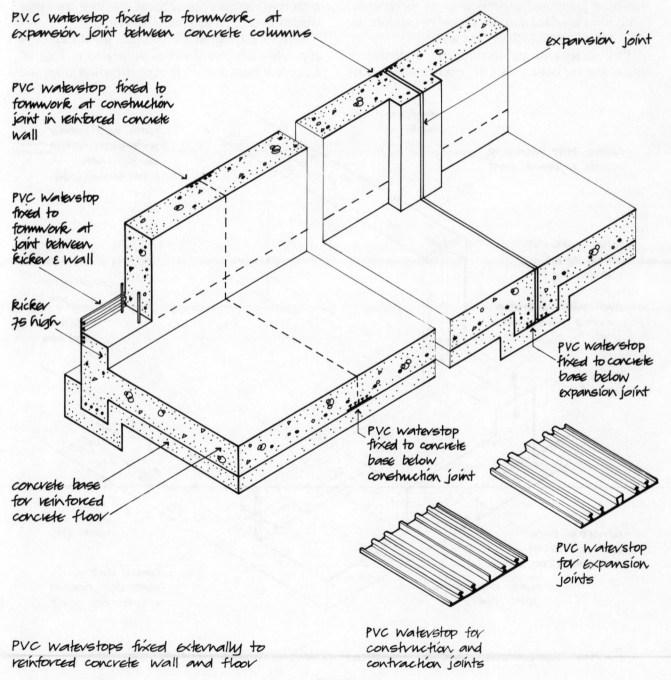

P.V.C waterstop fixed to formwork at expansion joint between concrete columns

PVC waterstop fixed to formwork at construction joint in reinforced concrete wall

PVC waterstop fixed to formwork at joint between kicker & wall

Kicker 75 high

concrete base for reinforced concrete floor

PVC waterstops fixed externally to reinforced concrete wall and floor

expansion joint

PVC waterstop fixed to concrete base below expansion joint

PVC waterstop fixed to concrete base below construction joint

PVC waterstop for construction and contraction joints

PVC waterstop for expansion joints

Fig. 27

between the bays. This procedure reduces overall shrinkage at the expense of an increase in construction joints which are in themselves a source of weakness.

Waterstops

As a barrier to penetration of water through construction joints and expansion joints in concrete, waterstops are fixed and cast against or cast into the thickness of concrete floors and walls.

PVC waterstops are cast into the underside of floors and the outside face of concrete walls across construction joints and expansion joints, as illustrated in Fig. 27. These flat faced waterstops are bonded to the concrete base under floors and fixed to the face of formwork for walls so that concrete is placed and compacted around the dumbell projections on the stops each side of the joint. The waterstop for expansion joints has a centre bulb that protrudes into the expansion joint to provide flexibility against movement. At the junction of joints, preformed cross-over sections of stop are heat welded to straight lengths of waterstop.

Rubber waterstops are cast into the thickness of concrete walls and floors as illustrated in Fig. 28. Plain web stops are cast in at construction joints and

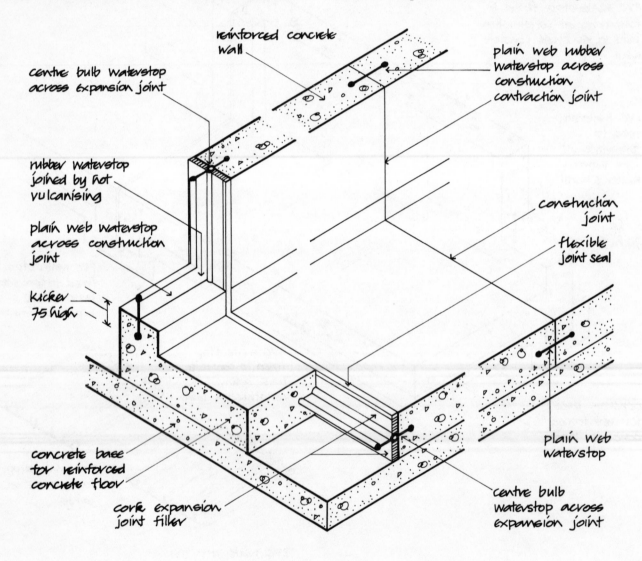

Rubber waterstops cast in reinforced concrete walls and floors

Fig. 28

26

centre bulb stops at expansion joints. These stops must be firmly fixed in place and supported with timber edging to one side of the stop so that concrete can be placed and compacted around the other half of the stop without moving it out of place. At the junction of joints the stops are joined by hot vulcanising.

For waterstops to be effective concrete must be placed and firmly compacted up to the stops and the stops must be secured in place to avoid them being displaced during placing and compacting of concrete. Waterstops will be effective in preventing penetration of water through joints providing they are solidly cast up to or inside sound concrete and there is no gross contraction at construction joints or movement at expansion joints.

Tanking

The term tanking is used to describe a continuous waterproof lining to the walls and floors of substructures to act as a tank to exclude water.

Mastic asphalt

The traditional material for tanking is mastic asphalt (see Chapter 4, Volume 1) which is applied and spread hot in three coats to a thickness of 20 for vertical and 30 for horizontal work. Joints between each laying of asphalt in each coat should be staggered at least 75 for vertical and 150 for horizontal work with the joints in succeeding coats. Angles are reinforced with a two coat fillet of asphalt.

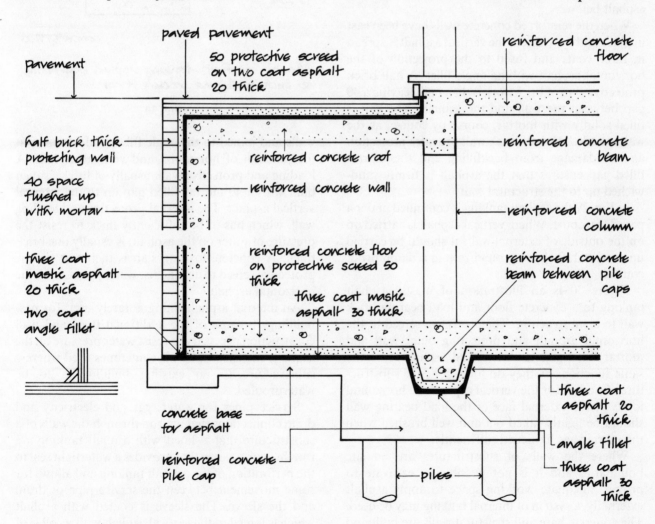

Mastic asphalt tanking applied externally to concrete wall and floor

Fig. 29

Asphalt tanking should be applied to the outside face of structural walls and under structural floors so that the walls and floors provide resistance against water pressure on the asphalt and the asphalt keeps water from the structure.

Figure 29 is an illustration of asphalt tanking applied externally to the reinforced concrete walls and floor of a substructure or basement. The horizontal asphalt is spread in three coats on the concrete base and over pile caps and extended 150 outside for the junction of the horizontal and vertical asphalt and the angle fillet. The horizontal asphalt is then covered with a protective screed of cement and sand 50 thick. The reinforced concrete floor should be cast on the protective screed as soon as possible to act as a loading coat against water pressure under the asphalt below.

When the reinforced concrete walls have been cast in place and have dried, the vertical asphalt is spread in three coats and fused to the projection of the horizontal asphalt with an angle fillet. A half brick protective skin of brickwork is then built leaving a 40 gap between the wall and the asphalt. The gap is filled solidly with mortar, course by course, as the wall is built. The half brick wall provides protection against damage from backfilling and the mortar filled gap ensures that the asphalt is firmly sandwiched up to the structural wall.

In Fig. 29 the asphalt tanking is continued under a paved forecourt. Where vertical asphalt is carried up on the outside of external walls it should be carried up at least 150 above ground to join a damp proof course.

Figure 30 is an illustration of mastic asphalt tanking to a concrete floor and load bearing brick wall to a substructure. The protective screed to the horizontal asphalt and protecting outer wall and mortar filled gap to the vertical asphalt serve the same functions as they do for a concrete substructure. As a key for the vertical asphalt the horizontal joints in the external face of the load bearing wall should be lightly raked out and well brushed when the mortar has hardened sufficiently.

Where the walls of substructures are on site boundaries and it is not possible to excavate to provide adequate working space to apply asphalt externally, a system of internal tanking may be used. The concrete base and structural walls are built and the horizontal asphalt is spread on the concrete base and a 50 protective screed spread over the asphalt. Asphalt is then spread up the inside of the structural

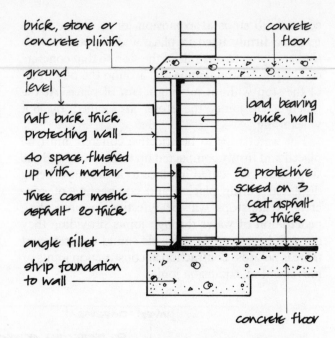

Mastic asphalt tanking applied externally to brick wall and concrete floor.

Fig. 30

walls and joined to the angle fillet reinforcement at the junction of horizontal and vertical asphalt. A loading and protective wall, usually of brick, is then built with a 40 mortar filled gap up to the internal vertical asphalt. The internal protective and loading wall, which has to be sufficiently thick to resist the pressure of water on the asphalt, is usually one brick thick. A concrete loading slab is then cast on the protective screed to act against water pressure on the horizontal asphalt.

An internal asphalt lining is rarely used for new buildings because of the additional floor and wall construction necessary to resist water pressure on the asphalt. Internal asphalt is sometimes used where a substructure to an existing building is to be waterproofed.

Service pipes for water, gas and electricity and drain connections that are run through the walls of a substructure that is lined with asphalt tanking are run through a sleeve that provides a watertight seal to the perforation of the asphalt tanking and allows for some movement between the service pipe or drain and the sleeve. The sleeve is coated with asphalt which is joined to the vertical asphalt with a collar of asphalt, run around the sleeve, as illustrated in Fig. 31.

Asphalt which is sandwiched in floors as tanking

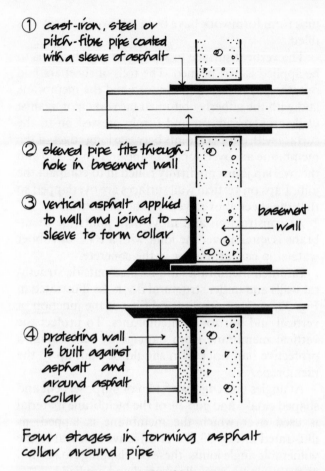

① cast-iron, steel or pitch-fibre pipe coated with a sleeve of asphalt

② sleeved pipe fits through hole in basement wall

③ vertical asphalt applied to wall and joined to sleeve to form collar → basement wall

④ protecting wall is built against asphalt and around asphalt collar

Four stages in forming asphalt collar around pipe

Fig. 31

has adequate compressive strength to sustain the loads normal to buildings. The disadvantages of asphalt are that asphalting is a comparatively expensive labour intensive operation and that asphalt is a brittle material that will readily crack and let in water if there is differential settlement or appreciable movements of the substructure. In general the use of asphalt tanking is limited to substructures with a length or width of not more than about seven and a half metres to minimise the possibility of settlement or movement cracks fracturing the asphalt.

Bituminous membranes

As an alternative to asphalt, bituminous membranes are commonly used for waterproofing and tanking to substructures. The membrane is supplied as a sheet of polythene or polyester film or sheet bonded to a self-adhesive rubber/bitumen compound or a polymer modified bitumen. The heavier grades of these membranes are reinforced with a meshed fabric sandwiched in the self-adhesive bitumen. The mem-

brane is supplied in rolls about one metre wide and twelve to eighteen metres long, with the self-adhesive surface protected with a release paper backing.

The particular advantage of these membranes is that their flexibility can accommodate small shrinkage, structural, thermal and moisture movements without damage to the membrane. Used in conjunction with waterstops to concrete substructures these membranes are generally preferable to asphalt as tanking.

The surface to which the membrane is applied by adhesion of the bitumen coating must be dry, clean and free from any visible projections that would puncture the membrane. The membrane is applied to a dry, clean float finished screed for floors and to level concrete wall surfaces on which all projecting

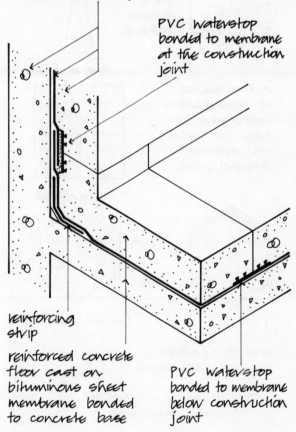

reinforced concrete wall cast against bituminous sheet membrane bonded to concrete retaining wall

PVC waterstop bonded to membrane at the construction joint

reinforcing strip

reinforced concrete floor cast on bituminous sheet membrane bonded to concrete base

PVC waterstop bonded to membrane below construction joint

Bituminous sheet membrane tanking to concrete

Fig. 32

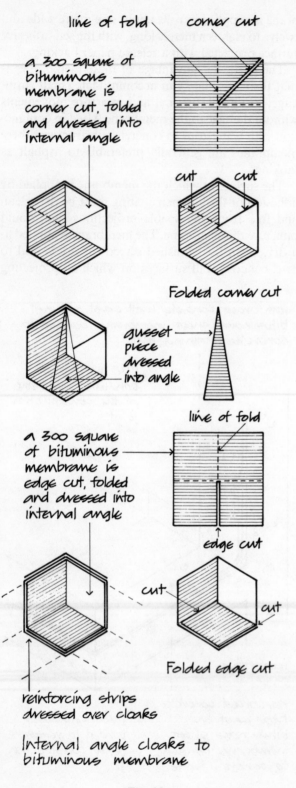

line of fold   corner cut

a 300 square of bituminous membrane is corner cut, folded and dressed into internal angle

cut   cut

Folded corner cut

gusset piece dressed into angle

line of fold

a 300 square of bituminous membrane is edge cut, folded and dressed into internal angle

edge cut

cut   cut

Folded edge cut

reinforcing strips dressed over cloaks

Internal angle cloaks to bituminous membrane

Fig. 33

nibs from formwork have been removed and cavities filled.

The vertical surface to which the membrane is to be applied is first primed. The rolls of sheet are laid out, the paper backing removed and the membrane laid with the adhesive bitumen face down or against walls and spread out and firmly pressed on to the surface with a roller. Joints between long edges of the membrane are overlapped 75 and end joints 150 and the overlap joints are firmly rolled in to compact the join. Laps on vertical wall surfaces are overlapped so that the sheet above overlaps the sheet below.

At construction and movement joints the membrane is spread over the joint with a PVC or rubber waterstop cast against or in the concrete.

Bitumen membranes are formed outside structural walls and under structural floors, as illustrated in Fig. 32, with an overlap and fillet at the junction of vertical and horizontal membranes. To protect the vertical membrane from damage by backfilling, a protective half brick skin should be built up to the membrane.

At angles and edges a system of purposely cut and shaped cloaks and gussets of the membrane material is used over which the membrane is lapped, as illustrated in Fig. 33. To be effective as a seal to the vulnerable angle joints, these overlapping cloaks and gusset must be carefully shaped and applied.

The effectiveness of these membranes as waterproof tanking depends on dry, clean surfaces free from protrusions or cavities and careful workmanship in spreading and lapping the sheets, cloaks and gussets.

CHAPTER TWO

STRUCTURAL STEEL FRAMES, FLOORS AND ROOFS

During the eighteenth century the application of steam power to large cast iron machinery, particularly in the textile grade, made it possible to gather large groups of labour together under one roof.

The earliest textile mills and warehouses were constructed with loadbearing brick or stone walls supporting timber floors and roofs. Heating was by open coal fires and lighting by oil lamps. Following a series of disastrous fires in mills, William Strutt designed and built in Derby the first fireproof mill in the years 1792–3. This six-storey mill consisted of loadbearing external walls and internal cast iron columns supporting timber beams at 3.0 centres between which brick arches were sprung. The arched brick floors were covered with sand on which tiles were laid.

A later development was the use of cast iron beams in place of timber. Up to the middle of the nineteenth century the majority of mills, factories and warehouses were constructed in this way and it was then that wrought iron began to replace cast iron as the material for beams. The failure of cast iron beams prompted the change. Following the invention in 1856 by Bessemer and in 1865 by Siemens and Martin of processes of converting iron into steel, mild steel began to compete with wrought iron and cast iron as a structural material. For many years, however, the Board of Trade limited the allowable stress in mild steel to that of wrought iron, that is 5 T/sq. in. (78 N/mm²), in tension and 4 T/sq. in. (63 N/mm²) in compression. Due to this it was generally cheaper to use wrought iron than steel up to 1897 when the Board of Trade relaxed its regulations to allow steel to be stressed at 8 T/sq. in. (125 N/mm²), the figure recommended by the British Association.

The first structural steel framed building to be erected in this country was the Ritz Hotel in London (1904–5) in which the whole of the weight of the masonry walls and floors and roofs was carried by the steel frame. A year later (1906) the east wing of Selfridges store was erected with a structural steel frame.

Up to the beginning of the Second World War (1939) the majority of tall buildings in this country were constructed with structural steel frames generally clad with brick or masonry to give the simulation of a large brick or masonry loadbearing structure.

The shortage of steel that followed the Second World War encouraged the use of reinforced concrete frames for buildings up to about 1980. Since 1980, due to considerable overproduction of steel and resulting competition, structural steel frames may be cheaper than comparable reinforced concrete frames.

The advantages of the structural steel frame are the speed of erection of the ready prepared steel members and the accuracy of setting out and connections that is a tradition in engineering works which facilitates accuracy in the fixing of cladding materials. With the use of sprayed on or dry lining materials to encase steel members to provide protection against damage by fire, a structural steel frame may be cheaper than a reinforced concrete structural frame because of speed of erection and economy in material and construction labour costs.

FUNCTIONAL REQUIREMENTS

The functional requirements of a structural frame are:

Strength and stability
Durability and freedom from maintenance
Fire safety

Strength and stability

The requirements from Part A of Schedule 1 to the Building Regulations 1991, as amended 1994, are that buildings be constructed so that the loadbearing elements, foundations, walls, floors and roofs have adequate strength and stability to support the dead loads of the construction and anticipated imposed loads on roofs, floors and walls without such undue deflection or deformation as might adversely affect the strength and stability of parts or the whole of the building. The strength of the loadbearing elements of the structure is assumed either from knowledge of the behaviour of similar traditional elements, such as walls and floors under load, or by calculations of the behaviour of parts or the whole of a structure under load, based on data from experimental tests, with various factors of safety to make allowance for unforeseen construction or design errors.

The strength of individual elements of a structure may be reasonably accurately assessed taking account of tests on materials and making allowance for variations of strength in both natural and man-made materials.

The strength of combinations of elements such as columns and beams depends on the rigidity of the connection and the consequent interaction of the elements. Here calculations make assumptions, based on tests, of the likely behaviour of the joined elements as a simple calculation or a more complex calculation of the behaviour of the parts of the whole of the structure. Various factors of safety are included in calculations to allow for unforeseen circumstances. Calculations of structural strength and stability provide a mathematical justification for an assumption of a minimum strength and stability of structures in use.

Imposed loads are those loads that it is assumed the building or structure is designed to support taking account of the expected occupation or use of the building or structure. Assumptions are made of the likely maximum loads that the floors of a category of building uses and may be expected to support. The load of the occupants and their furniture on the floors of residential buildings will generally be less than that of goods stored on a warehouse floor.

The loads imposed on roofs by snow are determined by taking account of expected snow loads in the geographical location of the building. Loads imposed on walls and roofs by wind (wind loads) are determined by reference to the situation of the building on a map of the United Kingdom on which basic wind speeds have been plotted. These basic wind speeds are the maximum gust speeds averaged over 3 second periods, which are likely to be exceeded on average only once in 50 years. In the calculation of the wind pressure on buildings a correction factor is used to take account of the shelter from wind afforded by obstructions and ground roughness (see Volume 2).

The stability of a building depends initially on a reasonably firm, stable foundation. The stability of a structure depends on the strength of the materials of the loadbearing elements in supporting, without undue deflection or deformation, both concentric and eccentric loads on vertical elements and the ability of the structure to resist lateral pressure of wind on walls and roofs.

The very considerable dead weight of walls of traditional masonry or brick construction is generally sufficient, by itself, to support concentric and eccentric loads and the lateral pressure of wind. The dead weight of skeleton framed multi-storey buildings is not generally, by itself, capable of resisting lateral wind pressure without undue deflection and deformation without some form of bracing to enhance stability. Unlike the joints in a reinforced concrete structural frame, the normal joints between vertical and horizontal members of a structural steel frame do not provide much stiffness in resisting lateral wind pressure.

Disproportionate collapse

A requirement from Part A of Schedule 1 to the Building Regulations 1991, as amended 1994, is that a building shall be constructed so that in the event of an accident, the building will not suffer collapse to an extent disproportionate to the cause. This requirement applies only to a building having five or more storeys (each basement level being counted as one storey) excluding a storey within the roof space where the slope of the roof does not exceed 70° to the horizontal.

The requirement to reduce the sensitivity of a building to disproportionate collapse in the event of an accident was included after the partial collapse of a multi-storey block of flats known as Ronan Point. This block of flats was constructed of precast reinforced concrete loadbearing, storey height panels supporting reinforced concrete floors. It appears that the occupant of a flat, on an intermediate floor, put a lighted match to her gas stove without realising that, due to an escape of gas, there had been a very considerable build-up of gas in her kitchen. There was an explosion that blew out wall panels and caused the partial collapse of floors above and below. On inspection it was determined that there was not sufficient horizontal tying of the floors to the wall panels to resist the unexpected and very considerable pressure of the explosion. Subsequently the building was demolished. Hence the new requirement to provide resistance to collapse disproportionate to the cause. The requirement is for the provision of horizontal tying adequate to a cause such as an explosion of this kind.

Durability and freedom from maintenance

The members of a structural steel frame are usually inside the wall fabric of buildings so that in usual circumstances the steel is in a comparatively dry atmosphere which is unlikely to cause progressive, destructive corrosion of steel. Structural steel will, therefore, provide reasonable durability for the expected life of the majority of buildings and require no maintenance.

Where the structural steel frame is partially or wholly built into the enclosing masonry or brick walls the external wall thickness is generally adequate to prevent such penetration of moisture as is likely to cause corrosion of steel. Where there is some likelihood of penetration of moisture to the structural steel, it is practice to provide protection by the application of paint or bitumen coatings or the application of a damp-proof layer. Where it is anticipated that moisture may cause corrosion of the steel, either externally or from a moisture laden interior, one of the weathering steels, that are much less subject to corrosion, is used.

Fire safety

The requirements from Part B of Schedule 1 to the Building Regulations 1991, as amended 1994, are concerned to ensure a reasonable standard of safety in case of fire. The application of the Regulations, as set out in the practical guidance given in Approved Document B, is directed to the safe escape of people from buildings in case of fire rather than the protection of the building and its contents. Insurance companies that provide cover against the risks of damage to the building and contents by fire may require additional fire protection such as sprinklers.

Internal fire spread (structures)

The requirement from the Regulations relevant to structure is to limit internal fire spread (structure).

As a measure of ability to withstand the effects of fire, the elements of a structure are given notional fire resistance times, in minutes, based on tests. Elements are tested for their ability to withstand the effects of fire in relation to:

(a) resistance to collapse (loadbearing capacity) which applies to loadbearing elements
(b) resistance to fire penetration (integrity) which applies to fire separating elements
(c) resistance to the transfer of excessive heat (insulation) which applies to fire separating elements.

The notional fire resisting times, which depend on the size, height and use of buildings, are chosen as being sufficient for the escape of occupants in the event of fire.

The requirements for the fire resistance of elements of a structure do not apply to:

(1) A structure that only supports a roof unless:
    (a) the roof acts as a floor, e.g. car parking, or as a means of escape and
    (b) the structure is essential for the stability of an external wall which needs to have fire resistance.
(2) The lowest floor of the building.

METHODS OF DESIGN

Permissible stress design method

With the introduction of steel as a structural material in the late nineteenth and early years of the twentieth century, the permissible stress method of design was accepted as a basis for the calculation of the sizes of structural members. Having established and agreed a yield stress for mild steel the permissible tensile stress

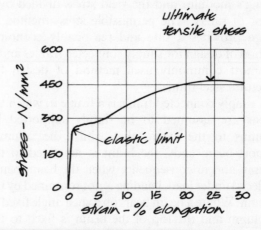

Stress/strain curve for mild steel

Fig. 34

was taken as the yield stress divided by a factor of safety to allow for unforeseen overloading, defective workmanship and variations in steel. The yield stress in steel is that stress at which the steel no longer behaves elastically and suffers irrecoverable elongation as shown in Fig. 34, which is a typical stress/strain curve for mild steel.

The loads to be carried by a structural steel frame are dead, imposed and wind loads. Dead loads comprise the weight of the structure including walls, floors, roof and all permanent fixtures. Imposed loads include all moveable items that are stored on or usually supported by floors, such as goods, people, furniture and moveable equipment. Wind loads are those applied by wind pressure or suction on the building. Dead loads can be accurately calculated. Imposed loads are assumed from the usual use of the building to give reasonable maximum loads that are likely to occur. Wind loads are derived from the maximum wind speeds.

Having determined the combination of loads that are likely to cause the worst working conditions the structure is to support, the forces acting on the structural members are calculated by the elastic method of analysis to predict the maximum elastic working stresses in the members of the structural frame. Beam sections are then selected so that the maximum predicted stress does not exceed the permissible stress.

In this calculation a factor of safety is applied to the stress in the material of the structural frame. The permissible compressive stress depends on whether a column fails due to buckling or yielding and is determined from the slenderness ratio of the column, Young's modulus and the yield stress divided by a factor of safety. The permissible stress method of design provides a safe and reasonably economic method of design for simply connected frames and is the most commonly used method of design for structural steel frames.

A simply connected frame is a frame in which the beams are assumed to be simply supported by columns to the extent that whilst the columns support beam ends, the beam is not fixed to the column and in consequence when the beam bends (deflects) under load, bending is not restrained by the column. Where a beam bears on a shelf angle fixed to a column and the top of the beam is fixed to the column by means of a small top cleat designed to maintain the beam in a vertical position, as illustrated later in Fig. 52, it is reasonable to assume that the beam is simply supported and will largely behave as if it had a pin jointed connection to the column.

Collapse or load factor method of design

Where beams are rigidly fixed to columns and where the horizontal or near horizontal members of a frame, such as the portal frame (see Volume 3), are rigidly fixed to posts or columns then beams do not suffer the same bending under load that they would if simply supported by columns or posts. The effect of the rigid connection of beam ends to columns is to restrain simple bending, as illustrated in Fig. 35. The fixed and beam bends in two directions, upwards near fixed ends and downwards at the centre. The upward bending is termed negative bending and the downward positive bending. It will be seen that bending at the ends of the beam is prevented by the rigid connections that take some of the stress due to loading and transfer it to the supporting columns. Just as the rigid connection of beam to column causes negative or upward bending of the beam at ends so a comparable, but smaller, deformation of the column will occur.

Using the elastic method of analysis to determine working stress in a fixed end beam to select a beam section adequate for the permissible stress design method, produces a section greater than is needed to

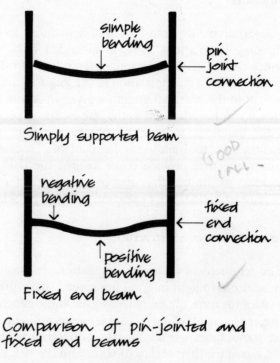

Comparison of pin-jointed and fixed end beams

Fig. 35

provide a reasonable factor of safety against collapse, because in practice the permissible stress is not reached and in consequence the beam could safely support a greater load.

The collapse or load factor method of design seeks to provide a load factor, that is a safety factor, against collapse applied to particular types of structural frame for economy in the use of materials by using the load factor which is applied to the loads instead of stress in materials.

The load factor method was developed principally for use in the design of reinforced concrete and welded connection steel frames with rigid connections as an alternative to the permissible stress method, as a means to economy in the selection of structural sections.

In the use of the load factor method of design plastic analysis is used. In this method of analysis of the forces acting in members it is presumed that extreme fibre stress will reach or exceed yield stress and the fibres behave plastically. This is a valid assumption as in practice the fibres of the whole section play a part in sustaining stress and under working loads extreme fibre stress would not reach yield point.

Limit state method of design

The purpose of structural analysis is to predict the conditions applicable to a structure that would cause it to become either unserviceable in use or unable to support loads to the extent that members might fail.

In the permissible stress method a limit is set on the predicted working stress in the members of the frame by the use of a factor of safety applied to the predicted yield stress of the materials used. In the load factor method of design a limit is set on the working loads to ensure that they do not exceed a limit determined by the application of a factor of safety to the loads that would cause collapse of the structure.

The limit state method of design seeks to determine the limiting states of both materials and loads that would cause a particular structure to become unserviceable in use or unsafe due to excessive load. The limiting conditions that are considered are serviceability during the useful life of the building and the ultimate limit state of strength.

Serviceability limit states set limits to the behaviour of the structure to limit excessive deflection, excessive vibration and irreparable damage due to material fatigue or corrosion that would otherwise make the building unserviceable in use.

Ultimate limit states of strength set limits to strength in resisting yielding, rupture, buckling and transformation into a mechanism and stability against overturning and fracture due to fatigue or low temperature brittleness.

In use the limit state method of design sets characteristic loads and characteristic strengths which are those loads and strengths that have an acceptable chance of not being exceeded during the life of the building. To take account of the variability of loads and strength of materials in actual use a number of partial safety factors may be applied to the characteristic loads and strengths, to determine safe working loads and strengths.

The limit state method of design has not been accepted wholeheartedly by structural engineers because, they say, it is academic, highly mathematical, increases design time and does not lead to economic structures. The opponents of limit state design prefer and use permissible stress design for simplicity in execution and the knowledge that the use of the more complex limit state design method may be rewarded with little significant reduction in frame sections.

Structural engineers profess to predict the likely behaviour of a structure from an acceptance of working loads and yield stresses in materials so as to design a structure that will be both safe and serviceable during the life of a building. There is often little reward in employing other than the permissible stress method of design for the majority of buildings so that the use of the limit state method is confined in the main to larger and more complex structures where the additional design time is justified by more adventurous and economic design.

STEEL SECTIONS

Mild steel

Mild steel is the material generally used for constructional steelwork. It is produced in several basic strength grades of which those designated as 43, 50 and 55 are most commonly used. The strength grades 43, 50 and 55 indicate minimum ultimate tensile strengths of 430, 500 and 550 N/mm^2 respectively.

Each strength grade has several subgrades indicated by a letter between A and E, the grades that are normally available are 43A, 43B, 43C, 43D, 43E, 50A, 50B, 50C, 50D and 55C. In each strength grade the subgrades have similar ultimate tensile strengths and as the subgrades change from A to E the specification becomes more stringent, the chemical composition changes and the notch ductility improves. The improvement in notch ductility (reduction in brittleness), particularly at low temperatures, assists in the design of welded connections and reduces the risk of brittle and fatigue failure which is of particular concern in structures subject to low temperatures.

Properties of mild steel

Strength

Steel is strong in both tension and compression with permitted working stresses of 165, 230 and 280 N/mm^2 for grades 43, 50 and 55 respectively. The strength to weight ratio of mild steel is good so that mild steel is able to sustain heavy loads with comparatively small self weight.

Elasticity

Under stress, induced by loads, a structural material will stretch or contract by elastic deformation and return to its former state once the load is removed. The ratio of stress to strain, which is known as Young's modulus (the modulus of elasticity), gives an indication of the resistance of the material to elastic deformation. If the modulus of elasticity is high the deformation under stress will be low. Steel has a high modulus of elasticity, 200 kN/mm^2, and is therefore a comparatively stiff material, which will suffer less elastic deformation than aluminium which has a modulus of elasticity of 69 kN/mm^2. Under stress, induced by loads, beams bend or deflect and in practice this deflection under load is limited to avoid cracking of materials fixed to beams. The sectional area of a mild steel beam can be less than that of other structural materials for given load, span and limit of deflection.

Ductility

Mild steel is a ductile material which is not brittle and can suffer strain beyond the elastic limit through what is known as plastic flow, which transfers stress to surrounding material so that at no point will stress failure in the material be reached. Because of the ductility of steel the plastic method of analysis can be used for structures with rigid connections, which makes allowance for transfer of stress by plastic flow and so results in a section less than would be determined by the elastic method of analysis, which does not make allowance for the ductility of steel.

Resistance to corrosion

Corrosion of steel occurs as a chemical reaction between iron, water and oxygen to form hydrated iron oxide, commonly known as rust. Because rust is open grained and porous a continuing reaction will cause progressive corrosion of steel. The chemical reaction that starts the process of corrosion of iron is affected by an electrical process through electrons liberated in the reaction, whereby small currents flow from the area of corrosion to unaffected areas and so spread the process of corrosion. In addition, pollutants in air accelerate corrosion as sulphur dioxides from industrial atmospheres and salt in marine atmospheres increase the electrical conductivity of water and so encourage corrosion. The continuing process of corrosion may eventually, over the course of several years, affect the strength of steel. Mild steel should therefore be given protection against corrosion in atmospheres likely to cause corrosion.

Fire protection

Although steel is non-combustible and does not contribute to fire it may lose strength when its temperature reaches a critical point in a fire in a building. A temperature of 550°C is generally accepted as the critical temperature for steel, which temperature will generally be reached in the early stages of a fire. To give protection against damage by fire, building regulations require fire protection of structural steelwork in certain situations.

Weathering steels

The addition of small quantities of certain elements modifies the structure of the rust layer that forms. The alloys encourage the formation of a dense fine grained rust film and also react chemically with sulphur in atmospheres to form insoluble basic sulphate salts which block the pores on the film and so prevent further rusting. The thin tightly adherent film that forms on this low alloy steel is of such low

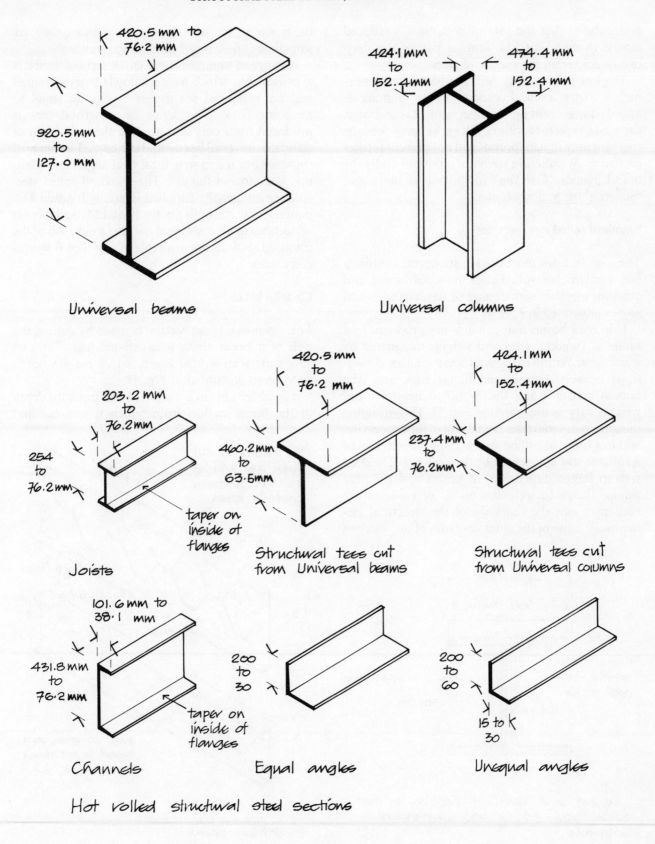

420.5 mm to
76.2 mm

920.5 mm
to
127.0 mm

Universal beams

424.1 mm
to
152.4mm

474.4 mm
to
152.4 mm

Universal columns

203.2 mm
to
76.2 mm

254
to
76.2 mm

Joists

420.5 mm
to
76.2 mm

460.2 mm
to
63.5 mm

taper on
inside of
flanges

Structural tees cut
from Universal beams

424.1 mm
to
152.4 mm

237.4 mm
to
76.2mm

Structural tees cut
from Universal columns

101.6 mm to
38.1 mm

431.8 mm
to
76.2 mm

taper on
inside of
flanges

Channels

200
to
30

Equal angles

200
to
60

15 to
30

Unequal angles

Hot rolled structural steel sections

Fig. 36

permeability that the rate of corrosion is reduced almost to zero. The film forms a patina of a deep brown colour on the surface of steel.

The low permeability rust film forms under normal wet/dry cyclical conditions. In conditions approaching constant wetness and in conditions exposed to severe marine or salt spray conditions the rust film may remain porous and not prevent further corrosion. Weathering steels are produced under the brand names 'Cor-Ten' for rolled sections and 'Stalcrest' for hollow sections.

Standard rolled steel sections

The steel sections most used in structural steelwork are standard hot rolled steel universal beams and columns together with a range of tees, channels and angles illustrated in Fig. 36.

Universal beams and columns are produced in a range of standard sizes and weights designated by serial sizes. Within each serial size the inside dimensions between flanges and flange edge and web remain constant and the overall dimensions and weights vary as illustrated in Fig. 37. This grouping of sections in serial sizes is convenient for production within a range of rolling sizes and for the selection of a suitable size and weight by the designer. The deep web to flange dimensions of beams and the near similar flange to web dimensions of columns are chosen to suit the functions of the structural elements. Because of the close similarity of the width of the flange to the web of column sections, they are sometimes known as 'broad flange sections'.

A range of comparatively small section 'joists' is also available, which have shallowly tapered flanges and are produced for use as beams for small to medium spans. The series of structural tees is produced from cuts that are half the web depth of standard universal beams and columns. The range of standard hot rolled structural steel angles and channels has tapered flanges. The standard rolled steel sections are usually supplied in strength grade 43A material with strength grades 50 and 55 available for all sections at an additional cost per tonne. All of the standard sections are available in Cor-Ten B weathering steel.

Castella beam

This open web beam section is made by cutting the web of a beam along a castellated line. The two halves are then welded together to form the castellated beam illustrated in Fig. 38.

The castella beam is one and a half times the depth of the beam section from which it was cut and

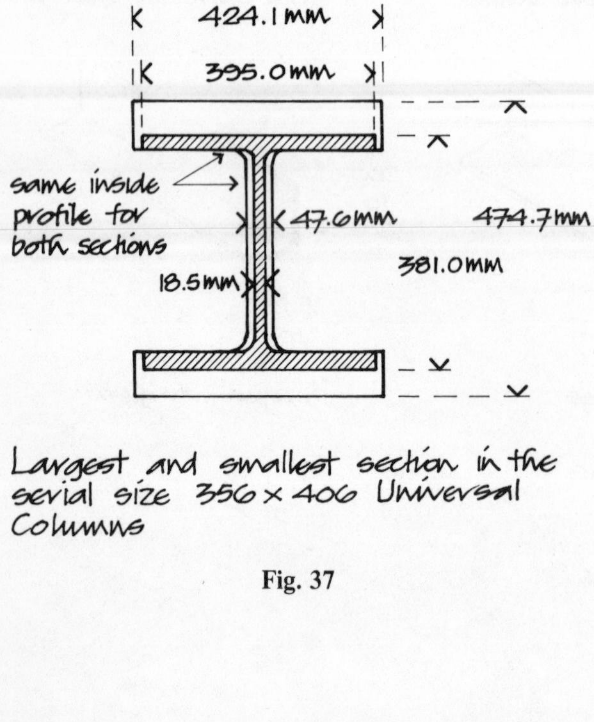

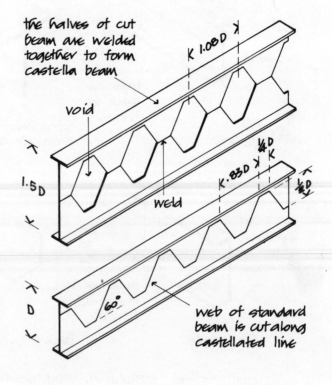

the halves of cut beam are welded together to form castella beam

void

weld

web of standard beam is cut along castellated line

60°

1.5D

D

K 1.08D

K .830D

¼D

½D

Castella beam

Fig. 38

same inside profile for both sections

424.1mm
395.0mm
47.6mm
18.5mm
474.7mm
381.0mm

Largest and smallest section in the serial size 356 × 406 Universal Columns

Fig. 37

therefore suffers less deflection under load. This section is economical for lightly loaded floors and long span roofs and the openings in the web are convenient for electrical and heating services.

Steel tubes

A range of seamless and welded seam steel tubes is manufactured for use as columns, struts and ties. The use of these tubes as columns is limited by the difficulty of making beam connections to a round section column. These round sections are the most efficient and compact structural sections available and are extensively used in the fabrication of lattice girders, columns, frames, roof decks and trusses for economy, appearance and comparative freedom from dust traps. Connections are generally made by scribing the ends of the tube to fit around the round sections to which they are welded. For long span members such as roof trusses, bolted plate connections are made at mid span for convenience in transporting and erecting long span members in sections.

Hollow rectangular and square sections

Hollow rectangular and square sections are made from round tube which, after heating, is passed through a series of rolls which progressively change the shape of the tube from a round to a square or rectangular section. To provide different wall thicknesses the tube can be reduced by stretching. The range of these sections is illustrated in Fig. 39. These sections are ideal for use as columns as the material is uniformly disposed around the axis and the rectangular section facilitates beam connections.

These hollow sections are used for lattice roof trusses and frames for the economy in material, particularly where the frame is exposed, and for the neat appearance of these sections which with welded connections have a more elegant appearance than angle sections. These sections are also much used in the fabrication of railings, balustrades, gates and fences.

Cold roll-formed steel sections

As long ago as 1936 a seven-storey block of flats was erected at Quarry Hill, Leeds, with a steel frame of cold rolled sections for both columns and beams. These sections are made from thin strips of steel

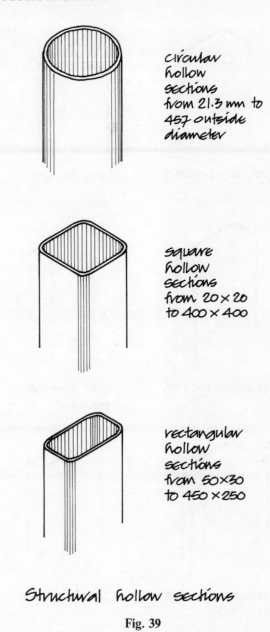

circular hollow sections from 21.3 mm to 457 outside diameter

square hollow sections from 20 × 20 to 400 × 400

rectangular hollow sections from 50×30 to 450 ×250

Structural hollow sections

Fig. 39

rolled to shape as illustrated in Fig. 40, which shows some of the common sections used. The advantage of cold rolling is that any shape can be produced to the exact dimensions to suit a particular use. Structural beam and column sections can be produced by welding sections together as illustrated in Fig. 41.

Cold rolled sections are extensively used in motor vehicles and domestic equipment, such as cookers, and as trim to a variety of cupboard fittings. In buildings, these sections are used for purlins, sheeting rails and cladding (see Volume 3). These sections are usually given a protective coating, such as galvanising or plastic coating, to inhibit corrosion.

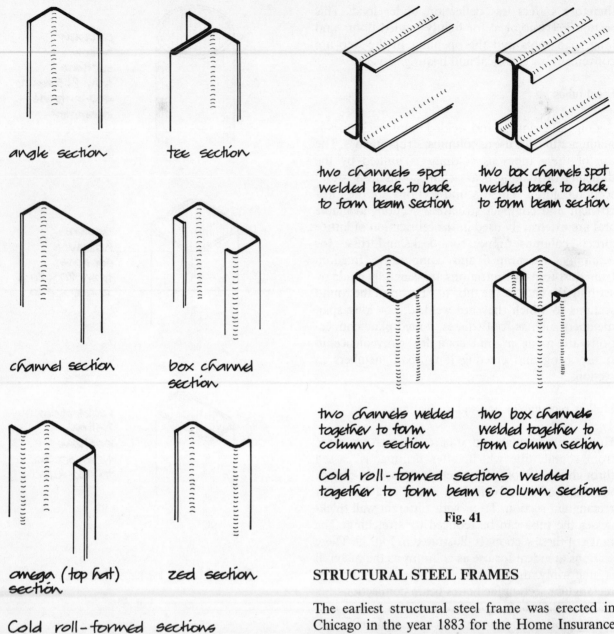

angle section   tee section

channel section   box channel
                       section

omega (top hat)   zed section
section

Cold roll-formed sections

Fig. 40

two channels spot   two box channels spot
welded back to back   welded back to back
to form beam section   to form beam section

two channels welded   two box channels
together to form   welded together to
column section   form column section

Cold roll-formed sections welded
together to form beam & column sections

Fig. 41

STRUCTURAL STEEL FRAMES

The earliest structural steel frame was erected in Chicago in the year 1883 for the Home Insurance building. At that time there were no limitations to the height of buildings, and the introduction of the passenger lift provided access to multi-storey buildings. Property taxes at that time were levied on site area so that developers were encouraged to obtain the maximum lettable floor area.

At the time the traditional method of building was solid load bearing walls of stone or brick. To use this system of building for multi-storey buildings would have necessitated walls of such thickness that there would have been an appreciable loss of floor area. The skeleton steel frame was introduced to reduce the thickness of external walls and so gain valuable

The advantage of these sections as a structural material is that the engineer can design the exact section he requires using the least material to achieve desired strength, so that with repetitive use of a few sections considerable economy of material is effected.

floor space. A skeleton of steel columns and beams carried the whole of the load of floors and the solid masonry or brick walls, the least thickness of which was dictated by weather resistance rather than load bearing requirements. Since then the steel frame has been one of the principal methods of constructing multi-storey buildings.

Skeleton frame

The conventional steel frame is constructed with hot rolled section beams and columns in the form of a skeleton designed to support the whole of the imposed and dead loads of floors, external walling or cladding and wind pressure. The arrangement of the columns is determined by the floor plans, horizontal and vertical circulation spaces and the requirements for natural light to penetrate the interior of the building.

Figure 42 is an illustration of a typical rectangular grid skeleton steel frame. In general the most economic arrangement of columns is on a regular rectangular grid with columns spaced at 3.0 to 4.0 apart parallel to the span of floors which bear on floor beams spanning up to 7.5, with floors designed to span one way between main beams. This arrange-

ment provides the smallest economic thickness of floor slab and least depth of floor beams, and therefore least height of building for a given clear height at each floor level.

Figure 43 is an illustration of a typical small skeleton steel frame designed to support one-way span floors on main beams and beams to support solid walls at each floor level on the external faces of the building. This rectangular grid can be extended in both directions to provide the required floor area.

Where it is inconvenient to have closely spaced internal columns, a larger rectangular or square grid is used as illustrated in Fig. 44, where each bay is divided by secondary beams spaced at up to 4.5 apart to carry one-way span floor slabs with the main floor beams which are supported by columns in turn supporting the secondary beams. This arrangement allows for the least thickness of floor slab, that is the least weight of construction. However, with increase in span of main floor beams goes increase in their depth and for a given minimum clear floor height between floor and soffit of beam this larger grid frame makes for greater overall height of building than does a smaller column grid.

Large floor areas unobstructed by columns can be framed by using deep long span solid web beams or by the use of deep lattice beams or a Vierendeel girder as illustrated in Fig. 45. The advantage of using a lattice or Vierendeel girder is that the girders may be designed so that their depth occupies the height of a floor and so does not increase the overall height of construction.

The conventional steel frame comprises continuous columns which support short beam lengths. This frame arrangement inhibits the use of projections from the rectangular frame because of the discontinuity of the beams which cannot be extended as cantilevers other than by the use of double columns. A solution is the use of continuous beams and short column lengths as illustrated in Fig. 46. This non-typical frame arrangement is practical and economic, particularly where welded built up beams and welded connections are used.

Wind bracing

The connections of beams to columns in multi-storey skeleton steel frames do not generally provide a sufficiently rigid connection to resist the considerable lateral wind forces that tend to cause the frame to rack. The word rack is used to describe the tendency

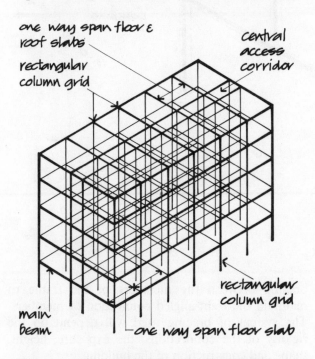

Rectangular grid structural steel frame

Fig. 42

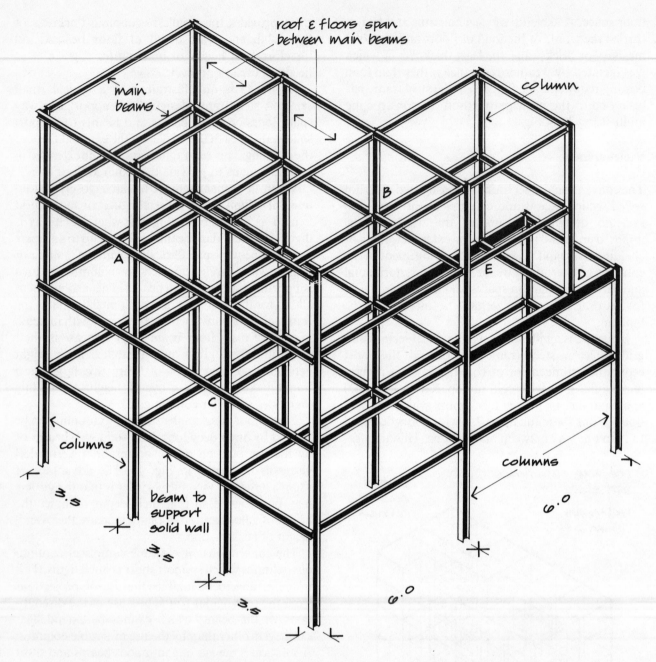

roof & floors span between main beams

main beams

column

columns

beam to support solid wall

columns

3.5

3.5

3.5

6.0

6.0

A

B

C

D

E

Structural steel skeleton frame

Fig. 43

of a frame to be distorted by lateral forces that cause right-angled connections to close up against the direction of the force in the same way that books on a shelf will tend to fall over if not firmly packed in place.

To resist racking caused by the very considerable wind forces acting on the faces of a multi-storey building it is necessary to include some system of

bracing between the members of the frame to maintain the right-angled connection of members. The system of bracing used will depend on the rigidity of the connections, the exposure, height, shape and construction of the building.

The frame for a 'point block' building, where the access and service core is in the centre of the building and the plan is square or near square, is commonly

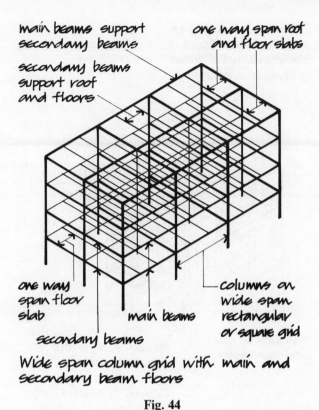

main beams support secondary beams

one way span roof and floor slabs

secondary beams support roof and floors

one way span floor slab

main beams

secondary beams

columns on wide span rectangular or square grid

Wide span column grid with main and secondary beam floors

Fig. 44

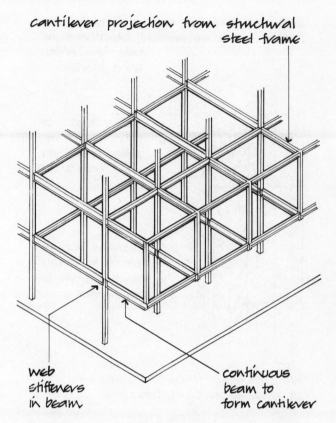

cantilever projection from structural steel frame

web stiffeners in beam

continuous beam to form cantilever

Continuous beam to form cantilever

Fig. 46

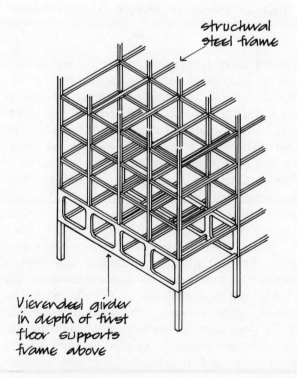

structural steel frame

Vierendeel girder in depth of first floor supports frame above

Vierendeel girder

Fig. 45

braced against lateral forces by connecting cross braces in the two sides of the steel frame around the centre core which are not required for access as illustrated in Fig. 47. Wind loads are transferred to the braced centre core through solid concrete floors acting as plates or by bracing steel framed floors.

With the access and service core on one face of the frame as illustrated in Fig. 48, the wind bracing can be connected in two opposite sides of the service core frame, leaving the other two sides for access and natural lighting. Wind forces are transferred to the braced service core by horizontal bracing to one or more of the framed floors.

With the skeleton steel frame to a 'slab block' which is rectangular on plan and has main facades much wider than end walls, it is common to connect cross braces to the end wall frames as illustrated in Fig. 49, to resist the racking effect of wind on the larger wall areas. Here it may be reasonable to accept that the wind forces acting on the smaller end walls will be resisted by the many connections of the two main wall frames and the horizontal plate floors.

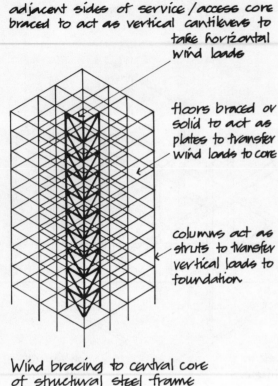

adjacent sides of service/access core braced to act as vertical cantilevers to take horizontal wind loads

floors braced or solid to act as plates to transfer wind loads to core

columns act as struts to transfer vertical loads to foundation

Wind bracing to central core of structural steel frame

Fig. 47

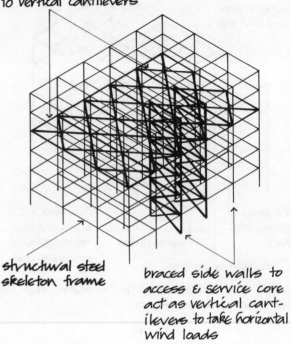

horizontal bracing to one or more floors transfers wind forces to vertical cantilevers

structural steel skeleton frame

braced side walls to access & service core act as vertical cant- ilevers to take horizontal wind loads

Wind bracing to service/access core and floor or floors of structural steel frame

Fig. 48

To provide fire protection to means of escape access and service cores to multi-storey buildings it is common to construct a solid reinforced concrete service core that by its construction and foundation acts as a stiff vertical cantilever capable of taking wind forces. Here wind forces are transferred to the core by solid plate floors, braced floors or by cantilever steel hangers that also serve to support floor beams as illustrated in Fig. 50.

Pin jointed structural steel frames

The shortage of materials and skilled craftsmen that followed the Second World War encouraged local authorities in this country to develop systems of building employing standardised components that culminated in the CLASP system of building.

The early development was carried out by the Hertfordshire County Council in 1945 in order to fulfil their school building programme. A system of prefabricated building components based on a square grid was developed, to utilise light engineering prefabrication techniques, aimed at economy by mass production and the reduction of site labour. Some ten years later the Nottinghamshire County

Council, faced with a similar problem and in addition the problem of designing a structure to accommodate subsidence due to mining operations, developed a system of building based on a pin jointed steel frame and prefabricated components. The pin jointed frame, with spring loaded diagonal braces, was designed to accommodate earth movements.

In order to gain the benefits of economy in mass production of component parts, the Nottinghamshire County Council joined with other local authorities to form CLASP (Consortium of Local Authorities Special Programme) which was able to order, well in advance, considerable quantities of standard components at reasonable cost. The CLASP system of building has since been used for schools, offices, housing and industrial buildings of up to four storeys. The system retained the pin jointed frame, originally designed for mining subsidence areas, as being the cheapest light structural steel frame.

The CLASP system is remarkable in that it was designed by architects for architects and allows a

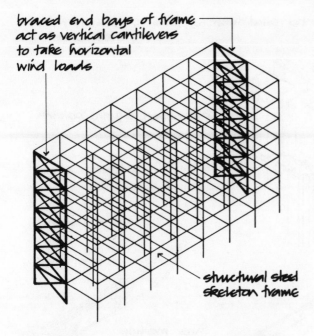

braced end bays of frame
act as vertical cantilevers
to take horizontal
wind loads

structural steel
skeleton frame

Wind bracing to end walls of structural
steel frame

Fig. 49

reinforced concrete core & steel cantilevers to hangers supporting floor beams

reinforced concrete
service/access core
takes horizontal
wind loads and
supports braced
steel cantilevers
that support
steel hangers
supporting floor
beams

Reinforced concrete core supporting
cantilever beams & steel hangers

Fig. 50

degree of freedom of design, within standard modules and using a variety of standard components, that no other system of prefabrication has yet to achieve. The CLASP building system is illustrated in Fig. 51.

Steel frame connections and fasteners

Connections between the members of a steel frame are made with angle section seating or shelf cleats to support beam end bearings on columns with top angle cleats, side cleats for beam to beam connections and splice connections between column lengths as illustrated in Fig. 52. These connections are made from short lengths of angle or tee section and plate that are fixed with bolt fasteners or by welding. To reduce on-site labour to a minimum, connecting cleats are fixed to columns and beams in the fabricator's shop as far as practicable for site assembly connections to be completed on site.

These simple cleat connections can be accurately and quickly made to provide support and connection between beams and columns. Cleat connections of beams to columns are generally assumed to provide a simple connection in structural analysis and calculation as there is little restraint to simple bending by the end connection of beams. This simply supported, that is unrestrained, connection is the usual basis for the design calculation for structural steel frames. Where the connection is made with high strength bolts or by welding it may be assumed that the connection is semi-rigid or rigid for design calculation purposes.

The connection of four beams to a column illustrated in Fig. 52, part B, shows a seating cleat shop bolted to the column to support the main beam and a side cleat bolted to the column to maintain the beam with its long axis vertical with the beam bolted to the cleats. The secondary beams bear on seating cleats bolted to the column with top angle cleats bolted to the beam and the column. This arrangement of connections of beams to columns was used where black bolts were used for site connections. Of recent years the use of shop welding has largely been abandoned in favour of the use of site bolted connections, using high strength friction bolts.

The connection of column lengths may be through plates to columns of similar section shown in Fig. 52, part C, or with a bearing plate, splice plates and packing pieces to connect columns of different section shown in Fig. 52, part F. In each case the

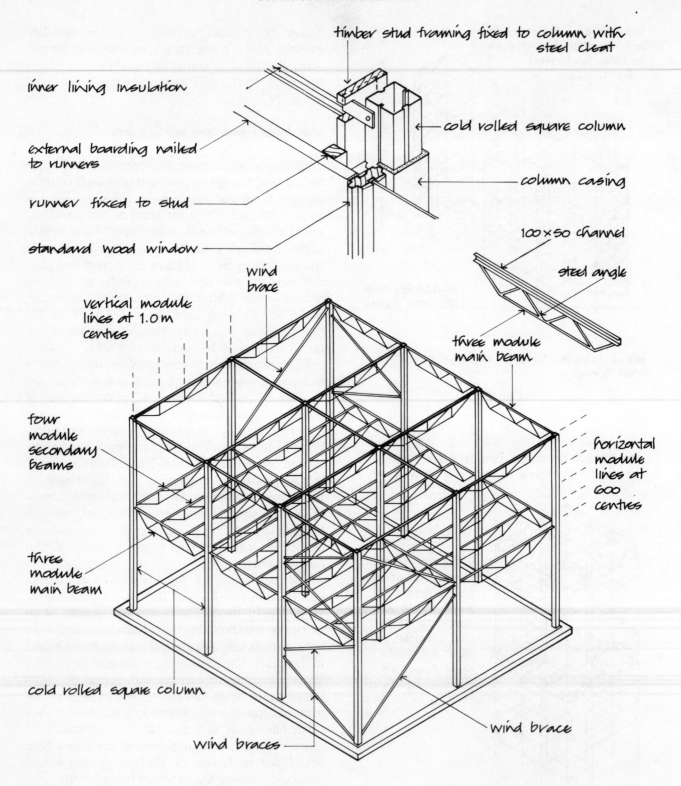

timber stud framing fixed to column with steel cleat

inner lining insulation

external boarding nailed to runners

runner fixed to stud

standard wood window

cold rolled square column

column casing

100 × 50 channel

steel angle

three module main beam

wind brace

vertical module lines at 1.0 m centres

four module secondary beams

horizontal module lines at 600 centres

three module main beam

cold rolled square column

wind brace

wind braces

Pin jointed steel frame

Fig. 51

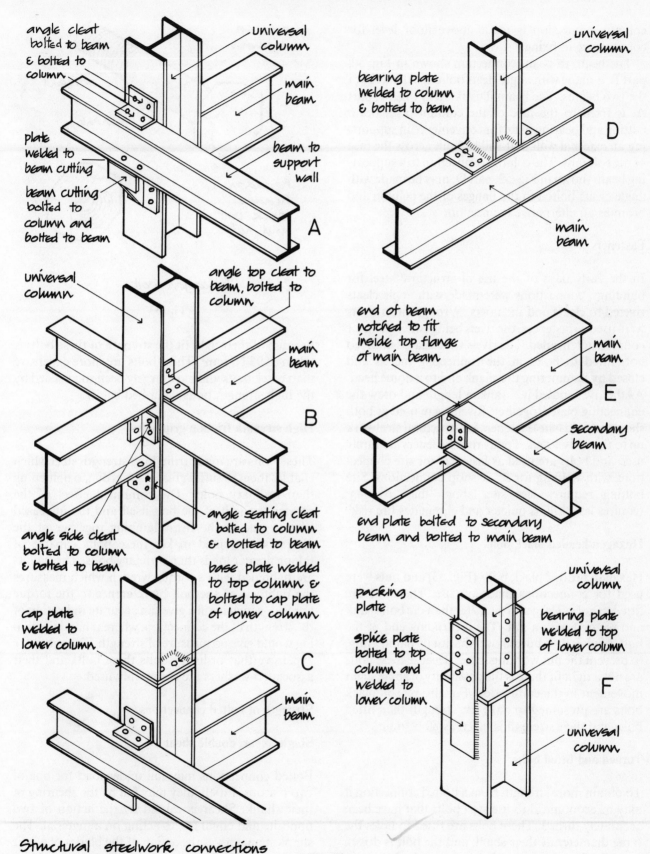

angle cleat
bolted to beam
& bolted to
column

universal
column

main
beam

plate
welded to
beam cutting

beam to
support
wall

beam cutting
bolted to
column and
bolted to beam

A

bearing plate
welded to column
& bolted to beam

universal
column

D

main
beam

universal
column

angle top cleat to
beam, bolted to
column

main
beam

B

end of beam
notched to fit
inside top flange
of main beam

main
beam

E

secondary
beam

angle side cleat
bolted to column
& bolted to beam

angle seating cleat
bolted to column
& bolted to beam

end plate bolted to secondary
beam and bolted to main beam

cap plate
welded to
lower column

base plate welded
to top column &
bolted to cap plate
of lower column

C

main
beam

packing
plate

universal
column

splice plate
bolted to top
column and
welded to
lower column

bearing plate
welded to top
of lower column

F

universal
column

Structural steelwork connections

Fig. 52

47

column connection is made above floor level for convenience in fixing.

The beam to beam connection shown in Fig. 52, part E, is made with angle cleats bolted to the webs of the two beams. The beam illustrated in Fig. 52, part A, is fixed to the face of the column on a beam cutting and bearing plate for convenience in supporting an external wall that will be built across the face of the columns. The column connection to a supporting beam shown in Fig. 52, part D, may be made with angle cleats bolted to the flanges of the column and beam as an alternative connection.

Fasteners

In the early days of the use of structural steel for buildings, connections were made with angle cleats riveted to cleats and members. Wrought iron rivets were used as fasteners, the rivets being either dome or countersunk headed. The rivets were heated until red hot, fitted to holes in the connecting metals and closed by hammering the shank end to a dome head. As the rivet cooled it shrank in length and drew the connecting plates together. Rivets were used as both shop and field (site) fasteners for structural steelwork up to the early 1950s. Today rivet fasteners are rarely used and bolts are used as fasteners for site connections with welding for some shop connections. Site bolting requires less site labour than riveting, requires less skill, is quieter and eliminates fire risk.

Hexagon headed black bolts

Hexagon headed black bolts (Fig. 53) and nuts were used for connections made on site. The bolts are fitted to holes 2 larger in diameter than the bolt shank and secured with a nut. The protruding end of the bolt shank is then burred over the nut by hammering, to prevent the nut working loose. Because these bolts are not a tight fit there is the possibility of some slight movement in the connection. For this reason black bolts are presumed to have less strength than fitted bolts and their strength is taken as 80 N/mm².

Turned and fitted bolts

To obtain more strength from a bolted connection it may be economical to use steel bolts that have been accurately turned. These bolts are fitted to holes the same diameter as their shank and the bolt is driven home by hammering and then secured with a nut.

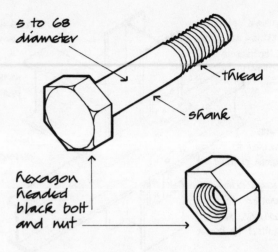

5 to 68 diameter

thread

shank

hexagon headed black bolt and nut

Black hexagon bolt

Fig. 53

Because of their tight fit the strength of these bolts is taken as 95 N/mm². These bolts are more expensive than black bolts and have largely been superseded by the high strength bolts described below.

High strength friction grip bolts

These bolts are made from high strength steel which enables them to suffer greater stress due to tightening than ordinary bolts. The combined effect of the greater strength of the bolt itself and the increased friction due to the firm clamping together of the plates being joined makes these bolts capable of taking greater loads than ordinary bolts. These bolts are tightened with a torque wrench which measures the tightness of the bolt by reference to the torque applied, which in turn gives an accurate indication of the strength of the connection, whereas hand tightening would give no measure of strength. Though more expensive that ordinary bolts these bolts and their associated washers are commonly used.

Strength of bolted connections

Single shear, double shear

Bolted connections may fail under load for one of two reasons. Firstly they may fail by the shearing of their shank. Shear is caused by the action of two opposite and equal forces acting on a material. The simplest analogy is the action of the blades of a pair of scissors or shears on a sheet of paper. As the blades close they exert equal and opposite forces which tear through the fibres of the paper forcing one part up and the other down. In the same way if the two plates joined by a bolt move with sufficient force in opposite directions than the bolt will fail in single shear as illustrated in Fig. 54. The strength of a bolt is determined by its resistance to shear in accordance with the strengths previously noted.

Where a bolt joins three plates it is liable to failure by the movement of adjacent plates in opposite directions as illustrated in Fig. 54. It will be seen that the failure is caused by the shank failing in shear at two points simultaneously, hence the term double shear. It is presumed that a bolt is twice as strong in double as in single shear.

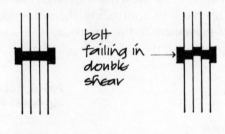

Double shear failure

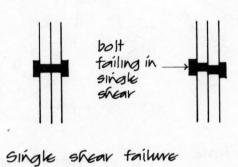

Single shear failure

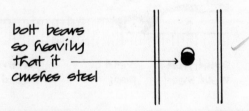

Bearing failure

Fig. 54

Bearing strength

The other type of failure that may occur at a connection is caused by the shank of the bolt bearing so heavily on the metal of the member or members it is joining that the metal becomes crushed as illustrated in Fig. 54. The strength of mild steel in resisting crushing due to the bearing of bolts is taken as 200 N/mm^2. The bearing area of a particular bolted connection is the product of the diameter of the bolt and the thickness of the thinnest member at the joint.

Bolt pitch (spacing)

If bolts are too closely spaced they may bear so heavily on the section of the members around them that they tear through the metal with the result that instead of the load being borne by all it may be transferred to a few bolts which may then fail in shear. To prevent the possibility of this type of failure it is usual to space bolts not less far apart than two and a half times their diameter, measured centre to centre and no bolt less than one and three quarter times its diameter from the edge of a member.

Welding

The word welding describes the operation of running molten weld metal into the heated junction of steel plates or members so that when the weld metal has cooled and solidified it strongly binds them together. The edges of the members to be joined are cleaned and also shaped for certain types of weld. For a short period the weld metal is molten as it runs into the joint, and for this reason it is obvious that a weld can be formed more readily with the operator working above the joint than in any other position. It will be seen that welding can be carried out more quickly and accurately in a workshop where the members can be manipulated more conveniently for welding than they can be on site.

Welding is most used in the prefabrication of built-up beams, trusses and lattice frames. The use of shop welded connections for angle cleats to conventional skeleton frames is less used than it was due to the possibility of damage to the protruding cleats during transport, lifting and handling of members.

In the design of welded structures it is practice to prefabricate as far as practical in the workshop and make site connections either by bolting or by designing joints that can readily be welded on site. The advantage of welding as applied to structural steel frames is that members can be built up to give the required strength for minimum weight of steel, whereas standard members do not always provide the most economical section. The labour cost in fabricating welded sections is such that it can only be justified in the main for long span and non-traditional frames. The reduction in weight of steel in welded frames may often justify higher labour costs in large heavily loaded structures. In buildings where the structural frame is partly or wholly exposed the neat appearance of the welded joints and connections is an advantage.

It is difficult to tell from a visual examination whether a weld has made a secure connection, and X-ray or sonic equipment is the only exact way of testing a weld for adequate bond between weld and parent metal. This equipment is somewhat bulky to use on site and this is one of the reasons why site welding is not favoured.

Surfaces to be welded must be clean and dry if the weld metal is to bond to the parent metal. These conditions are difficult to achieve in our wet climate out on site. The process of welding used in structural steelwork is 'fusion welding', in which the surface of the metal to be joined is raised to a plastic or liquid condition so that the molten weld metal fuses with the plastic or molten parent metal to form a solid weld or join.

For fusion welding the requirements are a heat source, usually electrical, to melt the metal, a consumable electrode to provide the weld metal to fill the gap between the members to be joined and some form of protection against the entry of atmospheric gases which can adversely affect the strength of the weld.

The metal of the members to be joined is described as the parent or base metal and the metal deposited from the consumable electrode, the weld metal. The fusion zone is the area of fusion of weld metal to parent metal.

The method of welding most used for structural steelwork is the arc welding process where an electric current is passed from a consumable electrode to the parent metals and back to the power source. The electric arc from the electrode to the parent metals generates sufficient heat to melt the weld metal and the parent metal to form a fusion weld.

The processes of welding most used are described in the following sections.

Manual metal-arc (MMA) welding

This manually operated process is the oldest and the most widely used process of arc welding. The equipment for MMA welding is simple and relatively inexpensive and the process is fully positional in that welding can be carried out vertically and even overhead due to the force with which the arc propels drops of weld metal on to the parent metal. Because of its adaptability this process is suitable for complex shapes, welds where access is difficult and on-site welding.

The equipment consists of a power supply and a hand held, flux covered, consumable electrode as illustrated in Fig. 55. As the electrode is held by hand the soundness of the weld depends largely on the skill of the operator in controlling the arc length and speed of movement of the electrode.

The purpose of the flux coating to the electrode is

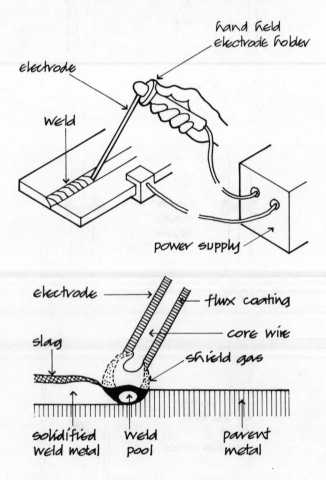

Manual metal arc welding

Fig. 55

to stabilise the arc, provide a gas envelope or shield around the weld to inhibit pick up of atmospheric gases and produce a slag over the weld metal to protect it from atmosphere.

Because this weld process depends on the skill of the operator there is a high potential for defects.

Metal inert-gas (MIG) and metal active-gas (MAG) welding

These processes use the same equipment, which is more complicated and expensive than that needed for MMA welding.

In this process the electrode is continuously fed with a bare wire electrode to provide weld metal and a cylinder to provide gas through an annulus to the electrode tip to form a gas shield around the weld as illustrated in Fig. 56. The advantage of the continuous electrode wire feed is that there is no break in welding to replace electrodes as there is with MMA welding, which can cause weakness in the weld run, and the continuous gas supply ensures a constant gas shield protection against the entry of atmospheric gases which could weaken the weld.

The manually operated electrode of this type of welding equipment can be used by less highly trained welders than the MMA electrode. The bulk of the

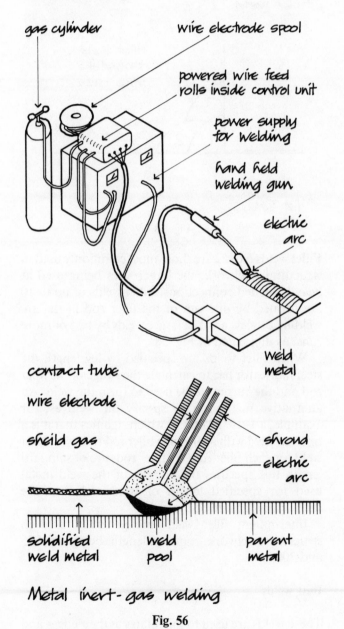

Metal inert-gas welding

Fig. 56

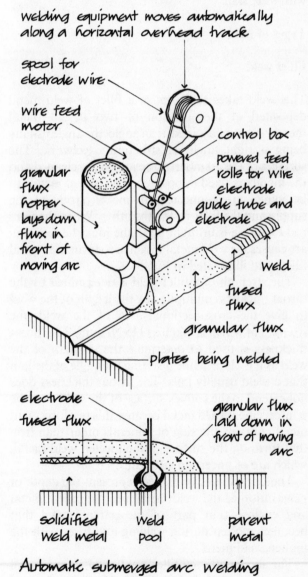

Automatic submerged arc welding

Fig. 57

51

equipment and the need for shelter to protect the gas envelope limit the use of this process to shop welding.

Submerged arc (SA) welding

Submerged arc (SA) welding is a fully automatic bare wire process of welding where the arc is shielded by a blanket of flux that is continuously fed from a hopper around the weld, as illustrated in Fig. 57. The equipment is mounted on a gantry that travels over the weld bench to lay down flux over the continuous weld run. The equipment, which is bulky and expensive, is used for long continuous shop weld runs of high quality that can be formed by welders with little skill.

Types of weld

Fillet weld

This weld takes the form of a fillet of weld metal deposited at the junction of two parent metal membranes to be joined at an angle, the angle usually being a right angle in structural steelwork. The surfaces of the members to be joined are cleaned and the members fixed in position. The parent metals to be joined are connected to one electrode of the supply and the filler rod to the other. When the filler rod electrode is brought up to the join, the resulting arc causes the weld metal to run in to form the typical fillet weld illustrated in Fig. 58.

The strength of a fillet weld is determined by the throat thickness multiplied by the length of the weld to give the cross-sectional area of the weld, the strength of which is taken as 115 N/mm^2. The throat thickness is used to determine the strength of the weld as it is along a line bisecting the angle of the join that a weld usually fails. The throat thickness does not extend to the convex surface of the weld over the reinforcement weld metal because this reinforcement metal contains the slag of minerals other than iron that form on the surface of the molten weld metal, which are of uncertain strength.

The dotted lines in Fig. 58 represent the depth of penetration of the weld metal into the parent metal and enclose that part of the parent metal that becomes molten during welding and fuses with the molten weld metal.

The leg lengths of fillet weld used in structural steelwork are 3, 4, 5, 6, 8, 10, 12, 15, 18, 20, 22, and 25, see Fig. 58. Throat thickness is leg length by 0.7.

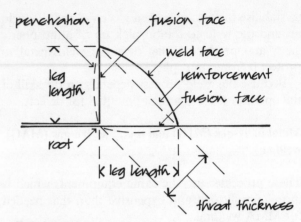

Fillet weld

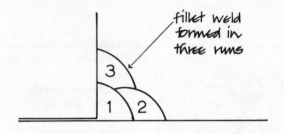

Fillet weld

Fig. 58

Fillet welds 5 to 22 are those most commonly used in structural steelwork, the larger sizes being used at heavily loaded connections. Fillet welds of up to 10 are formed by one run of the filler rod in the arc welding process and the larger welds by two or more runs, as illustrated in Fig. 58.

When fillet welds are specified by leg length the steel fabricator has to calculate the gauge of the filler rod and the current to be used to form the weld. An alternative method is to specify the weld as, for example, a 1–10/225 weld, which signifies that it is a one run weld with a 10 gauge filler rod to form 225 of weld for each filler rod. As filler rods are of standard length this specifies the volume of the weld metal used for specified length of weld and therefore determines the size of the weld.

Intermittent fillet welds are generally used in structural steelwork, common lengths being 150, 225 and 300.

Butt welds

These welds are used to join plates at their edges and

the weld metal fills the gap between them. The section of the butt weld employed depends on the thickness of the plates to be joined and whether welding can be executed from one side only or from both sides.

The edges of the plates to be joined are cleaned and shaped as necessary, the plates are fixed in position and the weld metal run in from the filler rod. Thin plates up to 5 thick require no shaping of their edges and the weld is formed as illustrated in Fig. 59. Plates up to 12 thick have their edges shaped to form a single V weld as illustrated in Fig. 60. The purpose of the V section is to allow the filler rod to be manipulated inside the V to deposit weld metal throughout the depth of the weld without difficulty.

Plates up to 24 thick are joined together either with a double V weld, where welding can be carried out from both sides, or by a single U where welding can only be carried out from one side. Figure 61 is an illustration of a double V and a single U weld. The U-shaped weld section provides room to manipulate the filler rod in the root of the weld but uses less of the expensive weld metal than would a single V weld of similar depth. It is more costly to form the edges of plates to the U-shaped weld than it is to form the V-shaped weld and the U-shaped weld uses less weld

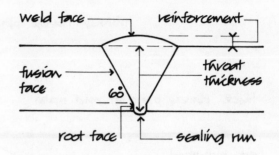

Single V butt weld

Fig. 60

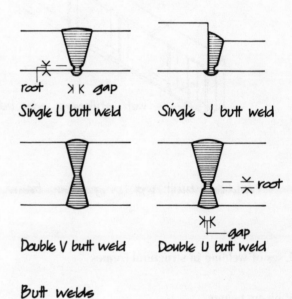

Double V butt weld   Double U butt weld

Butt welds

Fig. 61

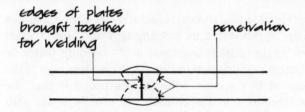

Deep penetration butt weld formed by welding from both sides

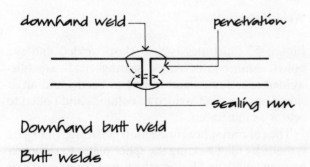

Downhand butt weld

Butt welds

Fig. 59

metal than does a V weld of similar depth. Here the designer has to choose the weld that will be the cheapest.

Plates over 24 thick are joined with a double U weld as illustrated in Fig. 61. Butt welds between plates of dissimilar thickness are illustrated in Fig. 61.

The throat thickness of a butt weld is equal to the thickness of the thinnest plate joined by the weld and the strength of the weld is determined by the throat thickness multiplied by the length of the weld to give the cross-sectional area of throat. The size of a butt weld is specified by the throat thickness, that is the thickness of the thinnest plate joined by the weld. The shape of the weld may be described in words as, for example, a double V butt weld or by symbols.

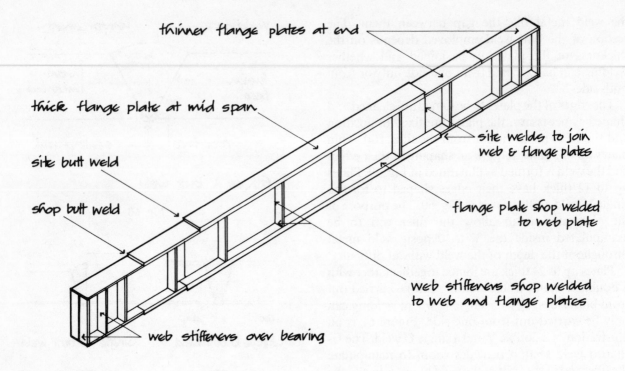

thinner flange plates at end

thick flange plate at mid span

site butt weld

shop butt weld

site welds to join web & flange plates

flange plate shop welded to web plate

web stiffeners shop welded to web and flange plates

web stiffeners over bearing

Welded built up long span beam

Fig. 62

Uses of welding in structural frames

Built-up beams

As has been stated, welding can often be used economically in fabricating large span beams whereas it is generally cheaper to use standard beam sections for medium and small spans.

Figure 62 is an illustration of a built-up beam section fabricated from mild steel strip and plates, fillet and butt welded together. It will be seen that the material can be disposed to give maximum thickness of flange plates at mid span where it is needed.

Figure 63 illustrates a welded beam end connection where strength is provided by increasing the size of the plates which are shaped for welding to the column.

Built-up columns

Columns particularly lend themselves to fabrication by welding where a fabricated column may be preferable to standard rolled steel sections. Columns can be fabricated by welding two standard channel or angle sections together, or by welding plates or beams as illustrated in Fig. 64. A disadvantage of the closed box section columns illustrated is that site connections to the column are usually made with long bolts passing right through the columns, which adds to the cost of the connection. This can be overcome by welding seat angles to the column and using a web cleat welded to the column in lieu of the conventional top angle cleat.

Welded connections

Figure 52 illustrates typical shop welded and site bolted connections. Angle seating cleats are fillet welded to columns and bolted to beams and angle side cleats are fillet welded to columns and bolted to beams as illustrated.

The column splice connection illustrated in Fig. 52 is made by fillet welding the splice plates to the lower column length and bolting the top of the splice plates through packing pieces.

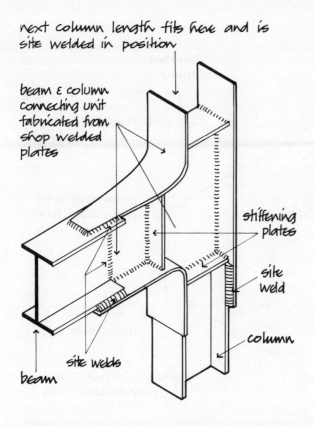

Welded beam to column connection

Fig. 63

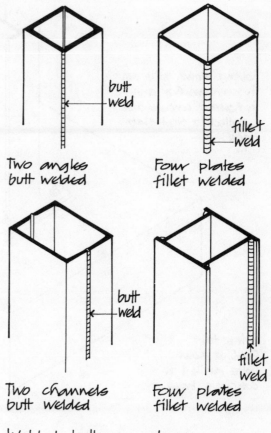

Welded butt-up columns

Fig. 64

Figure 65 illustrates a welded column connection directly to a slab base plate.

Column bases and foundations

Steel columns in framed buildings are used to support heavy loads and it is necessary to spread these loads from the comparatively small section of the columns to concrete, steel or reinforced concrete bases, which in turn spread the loads to the subsoil. The size of the foundation depends on the loads it supports and the bearing capacity of the subsoil. Where the subsoil for some depth below the surface has poor bearing capacity it is practice to use piles to transmit the loads from foundations down to a firm underlying stratum.

Column base plates

The bases of steel columns are accurately machined so that they bear truly on the steel base plates to which they are secured by welding. The purpose of

the base plate is to provide a sufficient area at the base of the column bearing on the concrete foundation that will not crush the concrete and to enable the base to be bolted to the concrete base. The two types of plate used are a comparatively thin plate 12 thick and the thicker bloom or slab illustrated in Fig. 65.

The base plate is either welded to the column or secured to it by means of gusset plates, as illustrated in Fig. 66.

Mass concrete foundation to columns

The base of columns carrying moderate loads of up to say 400 kN bearing on soils of good bearing capacity can be economically of mass concrete. The size of the base depends on the bearing capacity of the soil and load on the column base and the depth of the concrete is equal to the projection of the concrete beyond the base plate, assuming an angle of dispersion of load in concrete of 45 degrees. Figure 65 is an illustration of a mass concrete base.

Holding down bolts are either cast into the

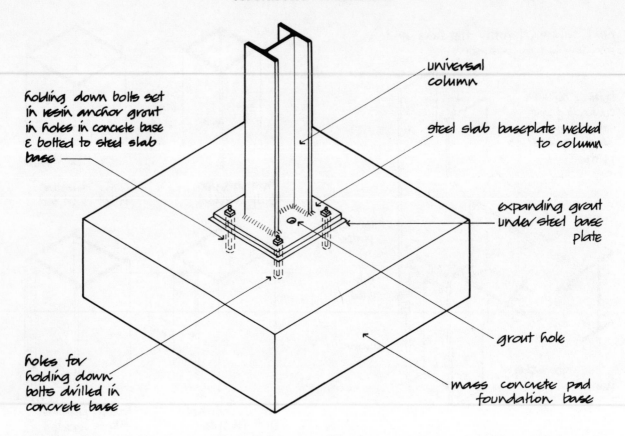

holding down bolts set in resin anchor grout in holes in concrete base & bolted to steel slab base

holes for holding down bolts drilled in concrete base

universal column

steel slab baseplate welded to column

expanding grout under steel base plate

grout hole

mass concrete pad foundation base

Steel slab base on concrete pad foundation

Fig. 65

concrete base or collars are cast in. When bolts are cast in concrete they are held in place whilst the concrete is being poured by means of a wood or metal template suspended over the concrete. It is often difficult to ensure that bolts are accurately cast in position on site and it is not uncommon for cast in bolts to be cut out and reset in position when the steel columns are being erected.

Collars are made of a sleeve of expanded steel as illustrated in Fig. 67 and the collars are cast into the concrete. The steel column is raised into position with holding down bolts in the collars. When the column has been levelled the holding down bolts are grouted into the collars. The advantages of these collars is that they allow room for the steel erector to manoeuvre the bolts until the column is in its correct position.

Columns are levelled in position on concrete bases by inserting steel wedges between the steel base plate and the concrete and these are adjusted until the column is level and plumb. Stiff concrete is then

hammered in between the steel plate and the concrete base to complete the foundation.

Reinforced concrete base

The area of the base required to spread the load from heavily loaded columns on subsoils of poor to moderate bearing capacity is such that it is generally more economical to use a reinforced concrete base than a mass concrete one. The steel column base plate is fixed as it is to a mass concrete base. Where column bases are large and closely spaced it is often economical to combine them in a continuous base or raft, as described in Chapter 1.

Steel grillage foundation

Steel grillage foundation is a base in which a grillage of steel beams transmits the column load to the subsoil. The base consists of two layers of steel beams, two or three in the top layer under the foot of

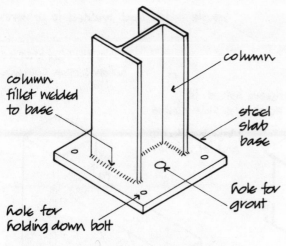

Column welded to slab base

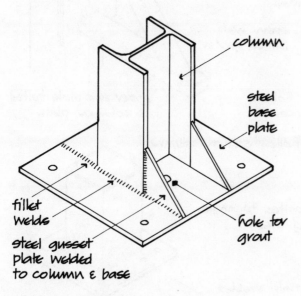

Column welded to plate base

Fig. 66

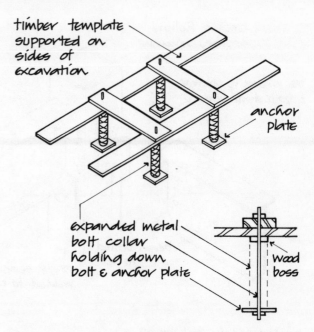

Expanded metal foundation bolt boxes

Fig. 67

the column and a lower cross layer of several beams so that the area covered by the lower layer is sufficient to spread column loads to the requisite area of subsoil. The whole of the steel beam grillage is encased in concrete. This type of base is rarely used as a reinforced concrete base is much cheaper.

Hollow rectangular sections

Beam to column connections

Bolted connections to closed box section columns

may be made with long bolts passing through the section. Long bolts are expensive and difficult to use as they necessitate raising beams on opposite sides of the column at the same time in order to position the bolts. Beam connections to hollow rectangular and square section columns may be made through plates, angles or tees welded to the columns. Standard beam sections are bolted to tee section cleats welded to columns and lattice beams by bolting end plates welded to beams to plates welded to columns, as illustrated in Fig. 68.

Flowdrill jointing

A recent innovation in making joints to hollow rectangular steel sections (HRS) is the use of the flowdrill technique as an alternative to the use of either long bolting through the section or site welding.

The flowdrill technique depends on the use of a tungsten carbide bit which can be used in a conventional drilling machine. As the tungsten carbide bit rotates on the surface of the HRS it generates sufficient heat to soften the steel so that the bit penetrates the steel wall of the HRS and redistributes the metal to form an internal bush as illustrated in

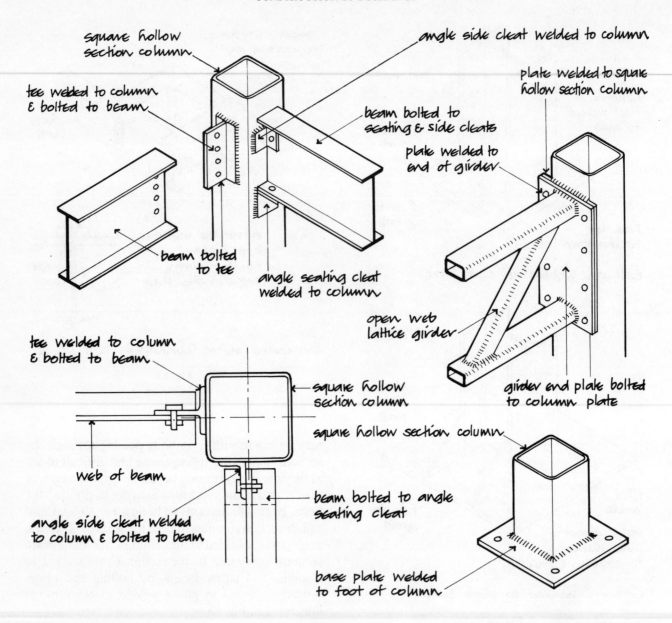

square hollow section column

angle side cleat welded to column

tee welded to column & bolted to beam

plate welded to square hollow section column

beam bolted to seating & side cleats

plate welded to end of girder

beam bolted to tee

angle seating cleat welded to column

open web lattice girder

girder end plate bolted to column plate

tee welded to column & bolted to beam

square hollow section column

square hollow section column

web of beam

beam bolted to angle seating cleat

angle side cleat welded to column & bolted to beam

base plate welded to foot of column

Connections to hollow section columns

Fig. 68

Fig. 69. Once the metal has cooled, the formed bush is threaded with a coldform flowtap to provide a threaded hole for a bolt. The beam connection to the HRS column is then completed by bolting end plates or web cleats to the beam and the ready drilled holes in the HRS column as illustrated in Fig. 69.

The flowdrill method of jointing is the preferred method of jointing for the benefit of economy in materials and site labour.

Cold strip sections

Beam to column connections

These connections are made by means of protruding studs or tees welded to the columns and bolted to the beams. Studs welded to columns are bolted to small section beams and ties and larger section beams to tee section cleats welded to columns, as illustrated in

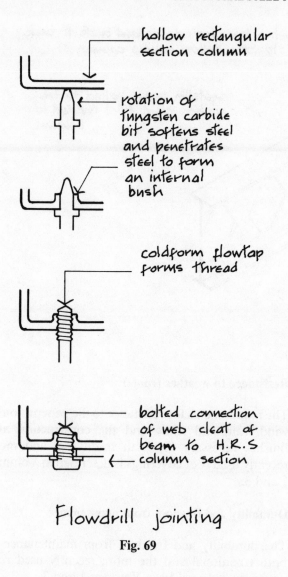

hollow rectangular section column

rotation of tungsten carbide bit softens steel and penetrates steel to form an internal bush

coldform flowtap forms thread

bolted connection of web cleat of beam to H.R.S column section

Flowdrill jointing

Fig. 69

Fig. 70. The tee section cleat is required for larger beams to spread the bearing area over a sufficient area of thin column wall to resist buckling.

Parallel beam structural steel frame

This type of structural steel frame uses double main or spine beams fixed each side of internal columns to support secondary rib beams that support the floor. The principal advantage of this form of structure is improved flexibility for the services, which can be located in both directions within the grid between the spine beams in one direction and the rib beams in the other. The advantage of using two parallel main spine beams is simplicity of connections to columns and the use of continuous long lengths of beam independent of column grid, which reduces fabrica-

tion and erection complexities and the overall weight of steel by the use of continuity of beams.

The most economical arrangement of the frame is a rectangular grid with the more lightly loaded rib beams spanning the greater distance between the more heavily loaded spine or main beams. Where long span ribs are used, for reasons of convenience in internal layout or for convenience in running services or both, a square grid may be most suitable.

The square grid illustrated in Fig. 71 uses double spine or main beams to internal columns with pairs of rib beams fixed to each side of columns with profiled steel decking and composite construction structural concrete topping fixed across the top of the rib beams. The spine beams are site bolted to end plates welded to short lengths of channel section steel that are shop welded to the columns. At the perimeter of the building a single spine beam is bolted to the end plate of channel sections welded to the column.

The parallel beam structural frame may be used, with standard I section beams and columns or with hollow rectangular section columns and light section rolled steel sections or cold formed strip steel beams and ribs, for smaller buildings supporting moderate floor loads in which there is need for provision for the full range of electric and electronic cables and air conditioning. Although the number of steel sections used for each grid of the framework in this system is greater than that needed for the conventional steel frame, there is generally some appreciable saving in the total weight and, therefore, the cost of the frame and appreciable saving in the erection time due to the simplicity of connections. The overall depth of the structural floor is greater than that of a similar conventional structural steel frame. As all the services common to modern buildings may be housed within the structural depth, rather than being slung below the structural floor of a conventional frame above a suspended ceiling, there may well be less overall height of building for given clear height between finished floor and ceiling level.

FLOORS AND ROOFS TO STRUCTURAL STEEL FRAMES

FUNCTIONAL REQUIREMENTS

The functional requirements of floors and roofs are:

Strength and stability
Resistance to weather

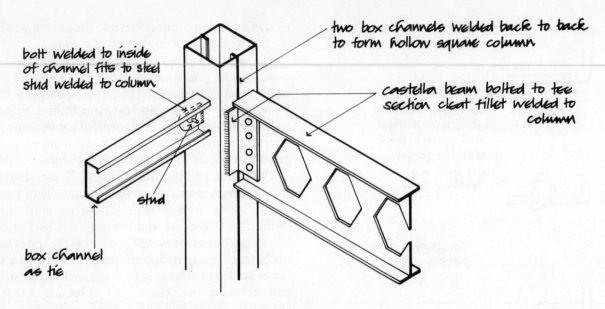

bolt welded to inside
of channel fits to steel
stud welded to column

two box channels welded back to back
to form hollow square column

castella beam bolted to tee
section cleat fillet welded to
column

stud

box channel
as tie

Cold roll-formed sections - connections

Fig. 70

Durability and freedom from maintenance
Fire safety
Resistance to the passage of heat
Resistance to the passage of sound

Strength and stability

The requirements from Part A of Schedule 1 to the Building Regulations 1991, as amended 1994, are that buildings be constructed so that the loadbearing elements, foundations, walls, floors and roofs have adequate strength and stability to support the dead loads of the construction and anticipated loads on roofs, floors and walls without such undue deflection or deformation as might adversely affect the strength and stability of parts or the whole of the building.

The strength and stability of floors and roofs depend on the nature of the materials used in the floor and roof elements and the section of the materials used in resisting deflection (bending) under the dead and imposed loads. Under load any horizontal element will deflect (bend) to an extent. Deflection under load is limited to about $\frac{1}{300}$ of span to minimise cracking of rigid finishes to floors and ceilings and to limit the sense of insecurity the occupants might have, were the floor to deflect too obviously. In general the strength and stability of a floor or roof is a product of the depth of the supporting members, the greater the depth the greater the strength and stability.

Resistance to weather (roofs)

The requirements for resistance to the penetration of wind, rain and snow and the construction and finishes necessary for both traditional and more recently used roof coverings is described in Volumes 1 and 3.

Durability and freedom from maintenance

The durability and freedom from maintenance of both traditional and the more recently used roof coverings is described in Volumes 1 and 3.

The durability and freedom from maintenance of floors constructed with steel beams, profiled steel decking and reinforced concrete depends on the internal conditions of the building. The majority of multi-storey framed buildings today are heated, so that there is little likelihood of moist internal conditions occurring, such as to cause progressive, destructive corrosion of steel during the useful life of the building.

Fire safety

The requirements from Part B of Schedule 1 to the Building Regulations 1991, as amended 1994, are concerned to:

(a) provide adequate means of escape
(b) limit internal fire spread (linings)

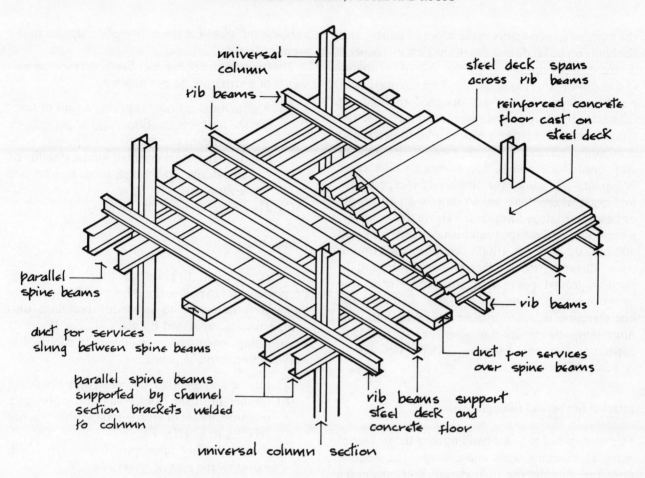

Parallel beam structural steel frame

Fig. 71

(c) limit internal fire spread (structure)
(d) limit external fire spread
(e) provide access and facilities for the Fire Services

Fire safety regulations are concerned to ensure a reasonable standard of safety in case of fire. The application of the regulations, as set out in the practical guidance given in Approved Document B, is directed to the safe escape of people from buildings in case of fire rather than the protection of the building and its contents. Insurance companies that provide cover against the risks of damage to the buildings and contents by fire will generally require additional fire protection such as sprinklers.

Means of escape

The requirement from the Building Regulations is that the building shall be designed and constructed so that there are means of escape from the building in case of fire to a place of safety outside the building. The main danger to people in buildings, in the early stages of a fire, is the smoke and noxious gases produced which cause most of the casualties and may also obscure the way to escape routes and exits. The Regulations are concerned to:

(a) provide a sufficient number and capacity of escape routes to a place of safety
(b) protect escape routes from the effects of fire by enclosure, where necessary, and to limit the ingress of smoke
(c) ensure the escape routes are adequately lit and exits suitably indicated

The general principle of means of escape is that any person in a building confronted by an outbreak of fire can turn away from it and make a safe escape. The number of escape routes and exits depends on the number of occupants in the room or storey, and the limits on travel distance to the nearest exit depend on the type of occupancy. The number of occupants in a room or storey is determined by the maximum number of people they are designed to hold, or calculated by using a floor space factor related to the type of accommodation which is used to determine occupancy related to floor area as set out in Approved Document B. The maximum number of occupants determines the number of escape routes and exits; where there are no more than 50 people one escape route is acceptable. Above that number, a minimum of 2 escape routes is necessary for up to 500 and up to 8 for 16 000 occupants. Maximum travel distances to the nearest exit are related to purpose-group types of occupation and whether one or more escape routes are available. Distances for one direction escape are from 9.0 to 18.0 and for more than one direction escape from 18.0 to 45.0, depending on the purpose groups defined in Approved Document B.

Internal fire spread (linings)

Fire may spread within a building over the surface of materials covering walls and ceilings. The Regulations prohibit the use of materials that encourage spread of flame across their surface when subject to intense radiant heat and those which give off appreciable heat when burning. Limits are set on the use of thermoplastic materials used in rooflights and lighting diffusers.

Internal fire spread (structure)

As a measure of ability to withstand the effects of fire, the elements of a structure are given notional fire resistance times, in minutes, based on tests. Elements are tested for the ability to withstand the effects of fire in relation to:

(a) resistance to collapse (loadbearing capacity) which applies to loadbearing elements
(b) resistance to fire penetration (integrity) which applies to fire separating elements
(c) resistance to the transfer of excessive heat (insulation) which applies to fire separating elements

The notional fire resistance times, which depend on the size, height and use of the building, are chosen as being sufficient for the escape of occupants in the event of fire.

The requirements for the fire resistance of elements of a structure do not apply to:

(1) A structure that only supports a roof unless:
   (a) the roof acts as a floor, e.g. car parking, or as a means of escape
   (b) the structure is essential for the stability of an external wall which needs to have fire resistance.
(2) The lowest floor of the building.

Compartments

To prevent rapid fire spread which could trap occupants, and to reduce the chances of fires growing large, it is necessary to subdivide buildings into compartments separated by walls and/or floors of fire-resisting construction. The degree of subdivision into compartments depends on:

(a) the use and fire load (contents) of the building
(b) the height of the floor of the top storey as a measure of ease of escape and the ability of fire services to be effective
(c) the availability of a sprinkler system which can slow the rate of growth of fire.

The necessary compartment walls and/or floors should be of solid construction sufficient to resist the penetration of fire for the stated notional period of time in minutes. The requirements for compartment walls and floors do not apply to single-storey buildings.

Concealed spaces

Smoke and flame may spread through concealed spaces, such as voids above suspended ceilings, roof spaces and enclosed ducts and wall cavities in the construction of a building. To restrict the unseen spread of smoke and flames through such spaces, cavity barriers and stops should be fixed as a tight fitting barrier to the spread of smoke and flames.

External fire spread

To limit the spread of fire between buildings, limits to the size of 'unprotected areas' of walls and also finishes to roofs, close to boundaries, are imposed by the Building Regulations. The term 'unprotected area' is used to include those parts of external walls that may contribute to the spread of fire between buildings. Windows are unprotected areas as glass offers negligible resistance to the spread of fire. The Regulations also limit the use of materials of roof coverings near a boundary that will not provide adequate protection against the spread of fire over their surfaces.

Access and facilities for the Fire Services

Buildings should be designed and constructed so that:

- Internal firefighting facilities are easily accessible
- Access to the building is simple
- Vehicular access is straightforward
- The provision of fire mains is adequate

Resistance to the passage of heat

The requirements for the conservation of power and fuel by the provision of adequate insulation of roofs are described in Volumes 1 and 3.

Resistance to the passage of sound

A description of the transmission and perception of sound is given in Volume 2. In multi-storey buildings the structural frame may provide a ready path for the transmission of impact sound over some considerable distance. The heavy slamming of a door, for example, can cause a sudden disturbing sound clearly heard some distance from the source of the sound by transmission through the frame members. Such unexpected sounds are often more disturbing than continuous background sounds such as eternal traffic noise. To provide resistance to the passage of such sounds it is necessary to provide a break in the path between potential sources of impact and continuous solid transmitters.

FLOOR AND ROOF CONSTRUCTION

In the early days of iron and steel framed construction, floors were constructed with comparatively closely spaced iron or steel filler beams, between main beams, giving support to shallow brick arches built between them on which concrete was spread to provide a level floor. This type of 'fire resisting' floor was in general use before the advent of reinforced concrete. The filler beam and brick arched floor gave way to the hollow clay block and in situ cast reinforced concrete floor described in Volume 1. This labour intensive form of construction was used for the advantage of the fire resisting property of the clay blocks and the reduction in dead weight afforded by the hollow blocks.

With the development of the technique of mass production of precast reinforced concrete units, systems of precast reinforced concrete slabs, beams and infill concrete block floors became common for steel framed buildings. The advantage of these floor systems was a considerable reduction in site labour and speed of erection because the slabs and beams were 'self-centering', that is they required no support from below once in place whereas previous floor systems required some support from below during construction.

Precast hollow floor beams

The precast, hollow, reinforced concrete floor units illustrated later in Fig. 111, Chapter 4, are from 400 to 1200 wide, 110 to 300 thick for spans of up to 10 metres for floors and $13\frac{1}{2}$ metres for the less heavily loaded roofs. The purpose of the voids in the units is to reduce deadweight without affecting strength. The reinforcement is cast into the webs between the hollows. The wide floor units are used where there is powered lifting equipment which can swing the units into place. These hollow floor units can be used by themselves as floor slabs with a non-structural levelling floor screed or they may be used with a structural reinforced concrete topping with tie bars over beams for composite action with the concrete casing to beams. End bearing of these units is a minimum of 75 on steel shelf angles or beams and 100 on masonry and brick walls.

The ends of these floor units are usually supported by steel shelf angles either welded or bolted to steel beams so that a part of the depth of the beam is inside the depth of the floor as illustrated later in Fig. 111. The ends of the floor units are splayed to fit under the top flange of the beams. A disadvantage of the construction shown later in Fig. 111 is that the deep I section beam projects some distance below the floor units and increases the overall height of construction for a given minimum clear height between floor and underside of beam. To minimise the overall height of

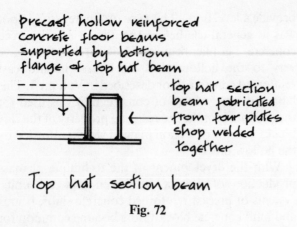

Precast hollow reinforced
concrete floor beams
supported by bottom
flange of top hat beam

top hat section
beam fabricated
from four plates
shop welded
together

Top hat section beam

Fig. 72

construction it is recent practice in Sweden to use welded top hat profile beams with the floor units supported by the bottom flange as illustrated in Fig. 72. The top hat section is preferred because of the difficulty of lowering and manoeuvring the units into the web of broad flange I section beams. This Swedish construction is particularly suited to multi-storey residential flats where the comparatively small imposed loads on floors facilitates a combination of overall beam depth and floor units to minimise construction depth. A screed is spread over the floor for lightly loaded floors and roofs and a reinforced concrete constructional topping for more heavily loaded floors.

Precast prestressed concrete floor units

These comparatively thin, prestressed solid plank, concrete floor units are designed as permanent centering (shuttering) for composite action with structural reinforced concrete topping as illustrated later in Fig. 112, Chapter 4.

The units are 400 and 1200 wide, 65, 75, or 100 thick and up to 9½ metres long for floors and 10 metres for roofs. It may be necessary to provide some temporary propping to the underside of these planks until the concrete topping has gained sufficient strength. A disadvantage of this construction is that as the planks are laid on top of the beams so that the floor spans continuously over beams, there is increase in overall depth of construction from top of floor to underside of beams.

Precast concrete tee beams

Precast concrete tee beam floors are mostly used for long span floors and particularly roofs of such

buildings as stores, supermarkets, swimming pools and multi-storey car parks where there is a need for wide span floors and roofs and the depth of the floor is no disadvantage. The floor units are cast in the form of a double tee as illustrated later in Fig. 113, Chapter 4. The strength of these units is in the depth of the tail of the tee which supports and acts with the comparatively thin top web. A structural reinforced concrete topping is cast on top of the floor units.

Precast beam and filler block floor

This floor system of precast reinforced concrete beams or planks to support precast hollow concrete filler blocks is illustrated later in Fig. 114, Chapter 4, for use with concrete beams. For use with steel beams the floor beams are laid between supports such as steel shelf angles fixed to the web of the beams or laid on the top flange of beams and the filler blocks are then laid between the floor beams. The reinforcement protruding from the top of the planks acts with the concrete topping to form a continuous floor system spanning across the structural beams. These small beams or planks and filler blocks can be manhandled into place without the need for heavy lifting equipment. This type of floor is most used in smaller scale buildings supporting the lighter imposed floor loads common in residential buildings for example.

Hollow clay block and concrete floor

This floor system, illustrated in Volume 1, consists of hollow clay blocks and in situ cast concrete reinforced as ribs between the blocks. This floor has to be laid on temporary centering to provide support until the in situ concrete has gained sufficient strength.

This floor system is much less used than it was due to the considerable labour required for setting up the temporary support and laying out the blocks and reinforcement.

Cold rolled steel deck and concrete floor

The traditional concrete floor to a structural steel frame consisted of reinforced concrete, cast in situ with the concrete casing to beams, cast on timber centering and falsework supported at each floor level until the concrete has sufficient strength to be self-supporting. The very considerable material and labour costs in erecting and striking the support for the concrete floor led to the adoption of the precast concrete self-centering systems such as the hollow beam and plank and beam and infill block floors. The term 'self-centering' derives from the word centering used to describe the temporary platform of wood or steel on which in situ cast concrete is formed. The precast concrete beam, plank and beam and block floors do not require temporary support, hence the term self-centering.

A disadvantage of the precast concrete beam and plank floors for use with a structural steel frame is that it is practice to erect the steel frame in one operation. Raising the heavy, long precast concrete floor units and moving them into position is to an extent impeded by the skeleton steel frame.

Of recent years profiled cold rolled steel decking, as permanent formwork acting as the whole or a part of the reinforcement to concrete, has become the principal floor system for structural steel frames. The profiled steel deck is easily handled and fixed in place as formwork (centering) for concrete.

The profiled cold roll-formed steel sheet decking, illustrated in Fig. 73, is galvanised both sides as a protection against corrosion. The profile is shaped for bond to concrete with projections that taper in from the top of the deck. Another profile is of trapezoidal section with chevron embossing for key to concrete.

The steel deck may be laid on the top flange of beams, as illustrated in Fig. 73, or supported by shelf

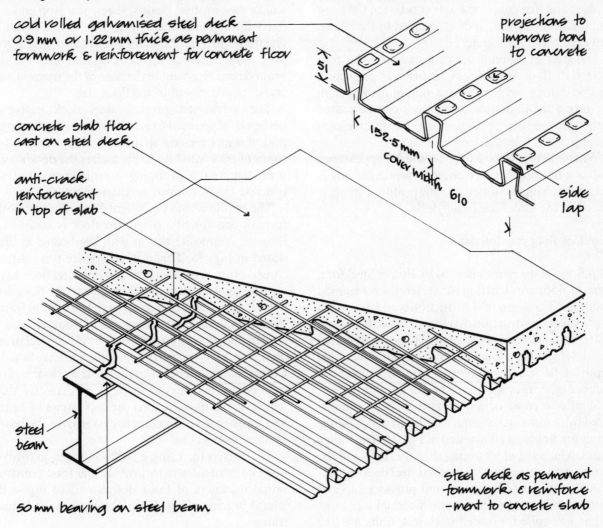

Cold rolled steel deck and concrete floor

Fig. 73

angles bolted to the web of the beam to reduce overall height and fixed in position on the steelwork with shot fired pins, self-tapping screws or by welding, with two fixings to each sheet. Side laps of deck are fixed at intervals of not more than one metre with self-tapping screws or welding.

For medium spans between structural steel beams the profiled steel deck acts as both permanent formwork and as reinforcement for the concrete slab that is cast in situ on the deck. A mesh of anti-crack reinforcement is cast into the upper section of the slab, as illustrated in Fig. 73.

For long spans and heavy loads the steel deck can be used with additional reinforcement cast into the bottom of the concrete between the upstanding profiles and, for composite action between the floor and the beams, shear studs are welded to the beams and cast into the concrete.

The steel mesh reinforcement cast into the concrete slab floor is sufficient to provide protection against damage by fire in most situations. For high fire rating the underside of the deck can be coated with sprayed on protection or an intumescent coating.

Where there is to be a flush ceiling for appearance and as a housing for services, a suspended ceiling is hung from hangers slotted into the profile or hangers bolted to the underside of the deck.

Slimfloor floor construction

'Slimfloor' is the name adopted by British Steel for a form of floor construction for skeleton steel framed buildings. This form of construction is an adaptation of a form of construction developed in Sweden where restrictions on the overall height of buildings dictated the development of a floor system with the least depth of floor construction to gain the maximum number of storeys within the height limitations.

Slimfloor construction comprises beams fabricated from universal column sections to which flange plates are welded as illustrated in Fig. 74. The flange plates, which are wider overall than the flanges of the beams, provide support to profiled steel decking that acts in part as reinforcement and provides support for the reinforced concrete constructional topping.

The galvanised, profiled steel deck units are 210 deep with ribs at 600 centres. The ribs and the top of the decking are ribbed to stiffen the plates and provide some bond to concrete. To seal the ends of the ribs in the decking to contain the concrete that

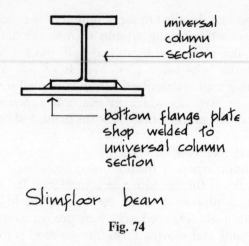

Slimfloor beam

Fig. 74

will be cast around beams, sheet steel stop ends are fixed through the decking to the flange plates as illustrated in Fig. 75. Constructional concrete topping is spread over the decking and into the ribs around reinforcement in the base of the ribs and anti-crack reinforcement in the floor slab.

The galvanised pressed steel deck units are designed for spans of 6 metres for use with the typical grid of 9 metre beam spans at 6 metre centres. For spans of over 6 and up to $7\frac{1}{2}$ metres the decking will need temporary propping at mid-span until the concrete has developed adequate strength.

The slimfloor may be designed as a non-composite form of construction where the floor is assumed to have no composite action with the beams as illustrated in Fig. 75. This non-composite type of floor construction is usual where the imposed floor loads are low, as in residential buildings, and the floor does not act as a form of bracing to the structural frame. Where buildings are in excess of four storeys in height and the imposed floor loads are relatively high, a composite form of construction may be used. Composite action between the concrete floor and the steel beams is achieved through 19 diameter studs which are shop-welded to the top flange of beams and transverse reinforcement cast in over the beams as illustrated in Fig. 76.

The reason for using a composite action form of floor construction is to provide the least constructional thickness of floor design and to utilise the lateral bracing effect of the floor on the structural frame.

A particular advantage of the slimfloor is that all or some of the various services, common to some modern buildings, may be accommodated within the deck depth rather than being slung below the

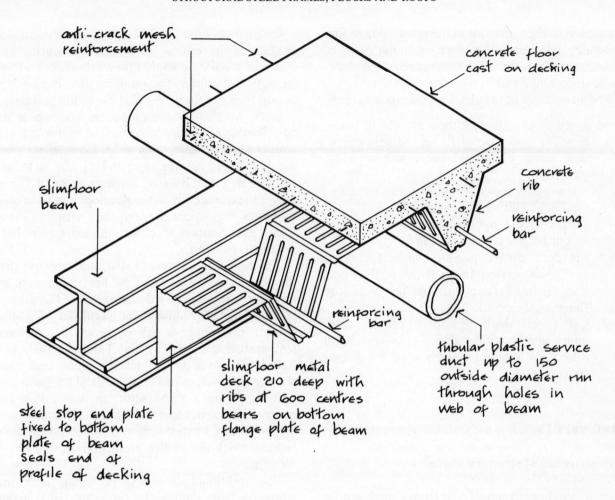

anti-crack mesh reinforcement

concrete floor cast on decking

slimfloor beam

concrete rib

reinforcing bar

reinforcing bar

tubular plastic service duct up to 150 outside diameter run through holes in web of beam

slimfloor metal deck 210 deep with ribs at 600 centres bears on bottom flange plate of beam

steel stop end plate fixed to bottom plate of beam seals end of profile of decking

Slimfloor construction

Fig. 75

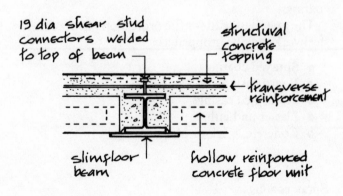

19 dia shear stud connectors welded to top of beam

structural concrete topping

transverse reinforcement

slimfloor beam

hollow reinforced concrete floor unit

Composite floor construction

Fig. 76

structural floor over a false ceiling. Calculations and tests have shown that 150 diameter holes may be cut centrally through the web of the beams at 600 spacing along the middle third of the length of the beam without significantly affecting the load carrying capacity of the beam. Figure 75 is an illustration of the floor system showing a plastic tube sleeve run through the web of a beam for service pipes and cables.

The ceiling finish may be fixed to the underside of the decking or hung from the decking to provide space for services such as ducting.

Because of the concrete encasement to the beams, most slimfloor constructions achieve 1 hour's fire resistance rating without the need for applied fire

67

protection to the underside of the beam. Where fire resistance requirement is over 60 minutes it is necessary to apply fire protection to the underside of the bottom flange plate.

The advantages of the slimfloor construction are:

(a) speed of construction through ease of man-handling and ease of fixing the lightweight deck units which provide a safe working platform

(b) pumping of concrete obviates the need for mechanical lifting equipment

(c) the floor slab is lightweight as compared to in situ or precast concrete floors

(d) the deck profile provides space for both horizontal services in the depth of the floor and vertical services through the wide top flange of the profile

(e) least overall depth of floor to provide minimum constructional depth consistent with robustness requirements dictated by design codes

FIRE SAFETY

Fire protection of structural steelwork

Fires in buildings generally start from a small source of ignition, the 'outbreak of fire', which leads to the 'spread of fire' followed by a steady state during which all combustible material burns steadily up to a final 'decay stage'.

Building regulations are mainly concerned with controlling the spread of fire to ensure the safety of those in the building and their safe escape in a notional period of time that varies from a half to six hours, depending on the use of the building, its construction and size.

To limit the growth and spread of fires in buildings the Regulations classify materials in accordance with the tendency of the materials to support spread of flame over their surface which is also an indication of the combustibility of the materials. Regulations also impose conditions to contain fires inside compartments to limit the spread of flame.

To provide safe means of escape, the Regulations set standards for the containment of fires and the associated smoke and fumes from escape routes for notional periods of time deemed adequate for escape from buildings.

One aspect of fire regulations is to specify notional periods of fire resistance for the loadbearing elements of a building so that they will maintain their strength and stability for a stated period during fires in buildings for the safety of those in the building.

Steel, which is non-combustible and makes no contribution to fire, loses so much of its strength at a temperature of 550°C that a loaded steel member would begin to deform, twist and sag and no longer support its load. Because a temperature of 550°C may be reached early in the development of fires in buildings, regulations may require a casing to structural steel members to reduce the amount of heat getting to the steel.

The larger the section of a structural steel member the less it will be affected by heat from fires by absorbing heat before it loses strength. The greater the mass and the smaller the perimeter of a steel section, the longer it will be before it reaches a temperature at which it will fail. This is due to the fact that larger sections will absorb more heat than smaller ones before reaching a critical temperature.

By applying a P/A factor, in which P is the perimeter in metres and A the cross-sectional area in square metres, to steel sections it is possible to use a reduced thickness of fire protection around heavy sections.

The traditional method of protecting structural steelwork from damage by fire is to cast concrete around beams and columns or to build brick or blockwork around columns with concrete casing to beams. These heavy, bulky and comparatively expensive casings have by and large been replaced by lightweight systems of fire protection employing sprays, boards, preformed casing and intumescent coatings.

The materials used for fire protection of structural steelwork may be grouped as:

- Sprayed coatings
- Board casings
- Preformed casings
- Plaster and lath
- Concrete, brick or block casings

Spray coatings

A wide range of products are available for application by spraying on the surface of structural steel sections to provide fire protection. The materials are

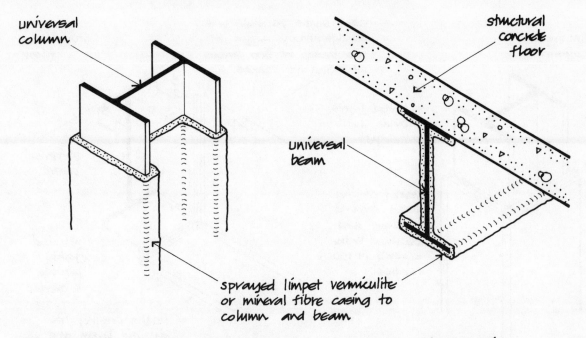

Fire protection of structural steelwork by sprayed limpet casing

Fig. 77

sprayed on to the surface of the steel sections so that the finished result is a lightweight coating that takes the profile of the coated steel, as illustrated in Fig. 77.

This is one of the cheapest methods of providing a fire protection coating or casing to steel for protection of up to four hours, depending on the thickness of the coating. The finished surface of these materials is generally coarse textured and because of the lightweight nature of the materials these coatings are easily damaged by knocks and abrasions. They provide some protection against corrosion of steel and, being lightweight, assist in controlling condensation.

These sprayed systems of protection are suitable for use where appearance is not a prime consideration and for beams in floors above suspended ceilings. Being lightweight and porous, spray coatings are not generally suited to external use.

Spray coatings may be divided into two broad groups as:

- Mineral fibre spray coatings
- Vermiculite/gypsum/cement spray coatings

Mineral fibre coatings

Mineral fibre coatings consist of mineral fibres that are mixed with inorganic binders, the wet mix being sprayed directly on to the clean, dry surface of the steel. The material dries to form a permanent, homogenous insulation that can be applied to any steel profile.

Vermiculite/gypsum/cement coatings

Vermiculite/gypsum/cement coatings consist of mixes of vermiculite or aerated magnesium oxychloride with cement or vermiculite with gypsum plaster. The materials are premixed and water is added on site for spray application directly to the clean, dry surface of steel. The mix dries to a hard, homogenous insulation that can be left rough textured from spraying or trowelled to a smooth finish. These materials are somewhat more robust than mineral spray coatings but will not withstand knocks.

Intumescent coatings

These coatings include mastics and paints which swell when heated to form an insulating protective coat which acts as a heat shield. The materials are applied by spray or trowel to form a thin coating over the profile of the steel section. They provide a hard finish which can be left textured from spraying or trowelled smooth, and provide protection of up to two hours.

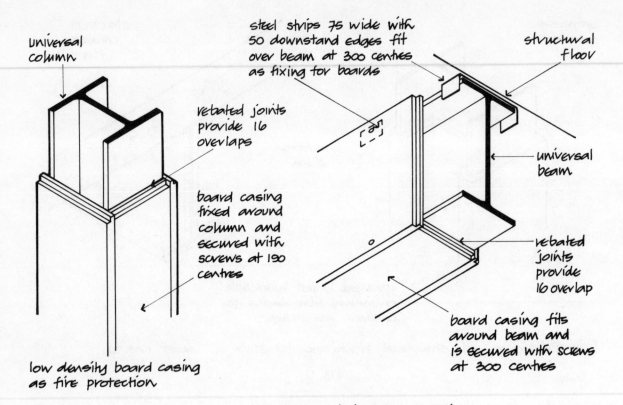

universal
column

steel strips 75 wide with
50 downstand edges fit
over beam at 300 centres
as fixing for boards

structural
floor

rebated joints
provide 16
overlaps

universal
beam

board casing
fixed around
column and
secured with
screws at 190
centres

rebated
joints
provide
16 overlap

low density board casing
as fire protection

board casing fits
around beam and
is secured with screws
at 300 centres

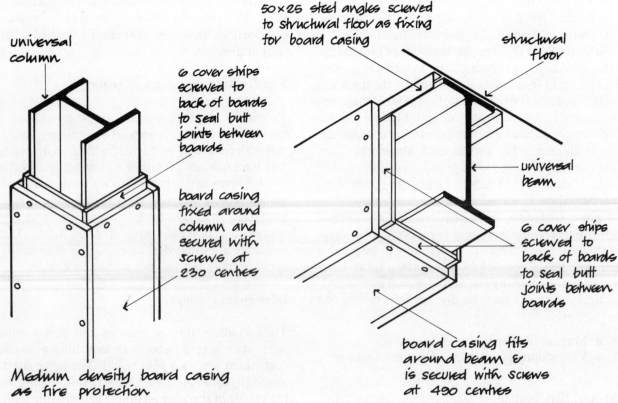

universal
column

50 × 25 steel angles screwed
to structural floor as fixing
for board casing

structural
floor

6 cover strips
screwed to
back of boards
to seal butt
joints between
boards

universal
beam

board casing
fixed around
column and
secured with
screws at
230 centres

6 cover strips
screwed to
back of boards
to seal butt
joints between
boards

Medium density board casing
as fire protection

board casing fits
around beam &
is secured with screws
at 490 centres

Board casing as fire protection for structural steelwork

Fig. 78

70

Board casings

There is a wide choice of systems based on the use of various preformed boards that are cut to size and fixed around steel sections as a hollow, insulating fire protection. Board casings may be grouped in relation to the materials that are used in the manufacture of the boards that are used as:

- Mineral fibre boards or batts
- Vermiculite/gypsum boards
- Plasterboard

For these board casings to be effective as fire protection they must be securely fixed around the steel sections, and joints between boards must be covered, lapped or filled to provide an effective seal to the joints in the board casing. These board casings,

which are only moderately robust, can suffer abrasion but are readily damaged by moderate knocks and are not suitable for external use. Board casings are particularly suitable for use in conjunction with ceiling and wall finishes of the same or like materials.

Mineral fibre boards and batts

Mineral fibre boards and batts are made of mineral fibres bound with calcium silicate or cement. The surface of the boards and batts, which is coarse textured, can be plastered. These comparatively thick boards are screwed to light steel framing around the steel sections. Mineral fibre batts are semi-rigid slabs which are fixed by means of spot-welded pins and lock washers. Mineral fibre boards

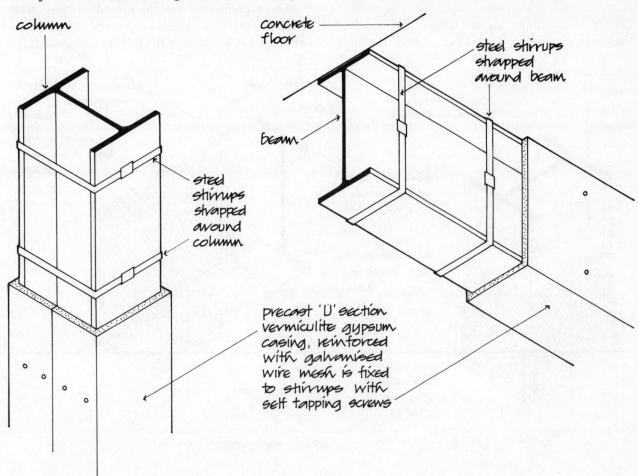

column

concrete floor

beam

steel stirrups strapped around beam

steel stirrups strapped around column

precast 'U' section vermiculite gypsum casing, reinforced with galvanised wire mesh is fixed to stirrups with self tapping screws

Fire protection of structural steelwork using precast vermiculite gypsum 'U' section casing

Fig. 79

71

are moderately robust and are used where appearance is not a prime consideration.

Vermiculite/gypsum boards

Vermiculite/gypsum boards are manufactured from exfoliated vermiculite and gypsum or non-combustible binders. The boards are cut to size and fixed around steelwork, either to timber noggins wedged inside the webs of beams and columns or screwed together and secured to steel angles or strips as illustrated in Fig. 78.

The edges of the boards may be square edged or rebated. The boards, which form a rigid, fairly robust casing to steelwork, can be self-finished or plastered.

Plasterboard casings

Plasterboard casings can be formed from standard thickness plasterboard or from a board with a gypsum/vermiculite core for improved fire resistance. The boards are cut to size and fixed to metal straps around steel sections. The boards may be self-finished or plastered. This is a moderately robust casing.

Preformed casings

These casings are made in preformed 'L' or 'U' shapes ready for fixing around the range of standard column or beam sections respectively. The boards are made of vermiculite and gypsum, or with a sheet steel finish on a fire resisting lining, as illustrated in Fig. 79. The vermiculite and gypsum boards are screwed to steel straps fixed around the steel sections and the sheet metal faced casings by interlocking joints and screws or by screwing.

These preformed casings provide a neat, ready finished surface with good resistance to knocks and abrasions in the case of the metal faced casings.

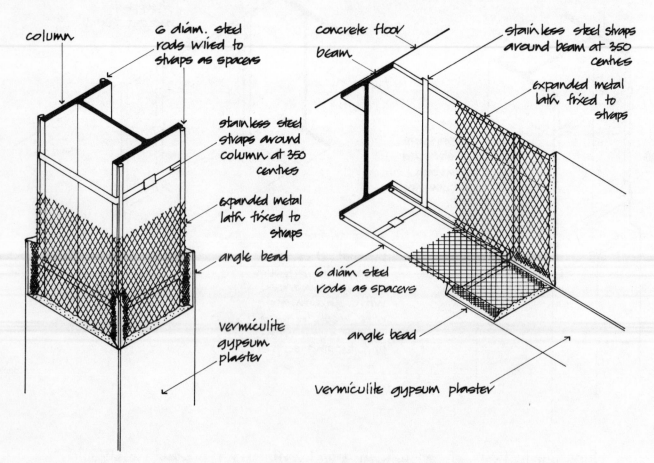

Fire protection of structural steelwork using metal lath and plaster casing

Fig. 80

Plaster and lath

Plaster on metal lath casing is one of the traditional methods of fire protection for structural steelwork. Expanded metal lath is stretched and fixed to stainless steel straps fixed around steel sections with metal angle beads at arrises, as illustrated in Fig. 80. The lath is covered with vermiculite gypsum plaster to provide an insulating fire protective casing that is trowelled smooth ready for decoration. This rigid, robust casing can suffer abrasion and knocks and is particularly suitable for use where a similar finish is used for ceilings and walls.

Concrete, brick or block casing

An in situ cast concrete casing is the traditional method of providing fire protection to structural steelwork and protection against corrosion. This solid casing is highly resistant to damage by knocks. To prevent the concrete spalling away from the steelwork during fires it is lightly reinforced, as illustrated in Fig. 81.

The disadvantages of a concrete casing to steelwork are its mass, which considerably increases the dead weight of the frame, and the cost of on site labour and materials in the formwork and falsework necessary to form and support the wet concrete.

Brick casings to steelwork may be used where brickwork cladding or brick division or compartment walls are a permanent part of the building, or where a brick casing is used for appearance sake to match surrounding fairface brick. Otherwise a brick

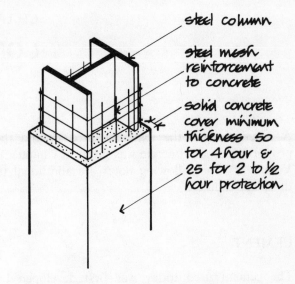

steel column

steel mesh reinforcement to concrete

solid concrete cover minimum thickness 50 for 4 hour & 25 for 2 to ½ hour protection

Non structural solid concrete fire protection to steel column

Fig. 81

casing is an expensive, labour intensive operation in the necessary cutting and bonding of brick around columns.

Blockwork may be used as an economic means of casing columns, particularly where blockwork divisions or walls are built up to structural steelwork. The labour in cutting and bonding these larger units is considerably less than with bricks. The blocks encasing steelwork are reinforced in every horizontal joint with steel mesh or expanded metal lath.

CHAPTER THREE
CONCRETE

A description of the materials used in and the proportioning of concrete was given in Chapter 1, Volume 1. The following notes are additional to those in Volume 1.

CEMENT

The cement used today was first developed by Joseph Aspdin, a Leeds builder, who took out a patent in 1824 for the manufacture of Portland cement. Aspdin developed the material for the production of artificial stone and named it Portland cement because, in its hardened state, it resembled natural Portland limestone in texture and colour. The materials of Aspdin's cement, limestone and clay, were later burned at a high temperature by Isaac Johnson in 1845 to produce a clinker which, ground to a fine powder, is what we now term Portland cement.

The characteristics of a cement depend on the proportions of the compounds of the raw materials used and the fineness of the grinding of the clinker, produced by burning the raw materials. A variety of Portland cements is produced, each with characteristics suited to a particular use.

The more commonly used Portland cements are:

- Ordinary Portland cement
- Rapid hardening Portland cement
- Sulphate resisting Portland cement
- White Portland cement
- Low heat Portland cement
- Portland blastfurnace cement
- Water repellent cement

Ordinary Portland cement

Ordinary Portland cement is the cheapest and most commonly used cement, accounting for about 90% of all cement production. It is made by heating limestone and clay to a temperature of about 1300°C to form a clinker, rich in calcium silicates. The clinker is ground to a fine powder with a small proportion of gypsum, which regulates the rate of setting when the cement is mixed with water. This type of cement is affected by sulphates such as those present in ground water in some clay soils. The sulphates have a disintegrating effect on ordinary Portland cement. For this reason sulphate-resisting cements are produced for use in concrete in sulphate-bearing soils, marine works, sewage installations and manufacturing processes where soluble salts are present.

Rapid hardening Portland cement

Rapid hardening Portland cement is similar to ordinary Portland except that the cement powder is more finely ground. The effect of the finer grinding is that the constituents of the cement powder react more quickly with water and the cement develops strength more rapidly.

Rapid hardening cement develops in three days, a strength which is similar to that developed by ordinary Portland in seven days. The advantage of the early strength developed by this cement is the possibility of speeding up construction by, for example, early removal of formwork. Although rapid hardening is more expensive than ordinary Portland cement, it is often used because of its early strength advantage. Rapid hardening Portland cement is not a quick setting cement. Several months after mixing there is little difference in the characteristics of ordinary and rapid hardening cements.

Sulphate resisting Portland cement

The proportions of the constituents of the cement that are affected by sulphates, that is aluminates, are reduced to provide increased resistance to the effect of sulphates. The effect of sulphates on ordinary cement is to combine with the constituents of the cement and the consequent increase in volume on crystallisation causes cement, and therefore concrete, to disintegrate. This disintegration is severe where the concrete is alternately wet and dry, as in marine works.

Because it is necessary to control, with some care, the composition of the raw materials of this cement it is more expensive than ordinary cement. High alumina cement described later is also a sulphate-resisting cement.

White Portland cement

White Portland cement is manufactured from china clay and pure chalk or limestone and is used to produce white concrete finishes. Due to the comparatively expensive raw material used (that is china clay) and the process of manufacture, it is considerably more expensive than ordinary cement and is used in the main for the surface of exposed concrete and for cement renderings. Pigments may be added to the cement to produce pastel colours.

Low heat Portland cement

Low heat Portland cement is used mainly for mass concrete works in dams and other constructions where the heat developed by hydration of other cements would cause serious shrinkage cracking. The heat developed by the hydration of cement in concrete in construction works is dissipated to the surrounding air, whereas in large mass concrete works it dissipates slowly. Control of the constituents of low heat Portland causes it to harden more slowly and therefore develop less rapidly than other cements. The slow rate of hardening does not affect the ultimate strength of the cement yet allows the low heat of hydration to dissipate through the mass of concrete to the surrounding air.

Portland blastfurnace cement

Portland blastfurnace cement is manufactured by grinding Portland cement clinker with blast furnace slag, the proportion of slag being up to 65% by weight and the percentage of cement clinker no less than 35%. This cement develops heat more slowly than ordinary cement and is used in mass concrete works as a low heat cement. It has good resistance to the destructive effects of sulphates and is commonly used in marine works.

Water repellent cement

Water repellent cement is made by mixing a metallic soap with ordinary or white Portland cement. Concrete made with this cement is more water repellent and therefore absorbs less rain water than concrete made with other cements and is thus less liable to dirt staining. This cement is used for cast concrete and cast stone for its water repellent property.

High alumina (aluminous) cement

High alumina (aluminous) cement is not one of the Portland cements. It is manufactured from bauxite and limestone or chalk in equal proportions. Bauxite is a mineral containing a higher proportion of alumina (aluminium oxide) than the clays used in the manufacture of Portland cements, hence the name given to this cement.

The disadvantages of this cement are that there is a serious falling off in strength in hot moist atmospheres, and it is attacked by alkalis. This cement is no longer used for concrete.

AGGREGATES

Concrete is a mix of particles of hard material, the aggregate, bound with a paste of cement and water with at least three quarters of the volume of concrete being occupied by aggregate. Volume for volume, cement is generally more costly than aggregate and it is advantageous, therefore, to use as little cement as necessary to produce a dense, durable concrete.

A wet concrete mix is spread in the form of foundation bases, slabs or inside formwork for beams and columns and compacted into a dense mass. There is a direct relation between the density and strength of finished concrete and the ease with which concrete can be compacted. The characteristics of the aggregate play a considerable part in the ease with which concrete can be compacted. The measure of the ease with which concrete can be compacted is described as the workability of the mix. Workability is affected by the characteristics of the particles of aggregate such as size and shape, so that for a given mix workability can be improved by careful selection of aggregate.

The grading of the size and the shape of the particles of aggregate affects the amount of cement and water required to produce a mix of concrete that is sufficiently workable to be compacted to a dense mass. The more cement and water that are needed for the sake of workability, the greater the drying shrinkage there will be by loss of water as the concrete dries and hardens.

Characteristics of aggregate

Aggregate for concrete should be hard, durable and contain no materials that are likely to decompose or change in volume or affect reinforcement. Clay, coal or pyrites in aggregate may soften, swell, decompose and cause stains in concrete.

Aggregate should be clean and free from organic impurities and coatings of dust or clay that would prevent the particles of aggregate from being adequately coated with cement and so lower the strength of the concrete.

Types of aggregate

Natural aggregates

Sand and gravel are the cheapest and most commonly used aggregate in this country and consist of particles of broken stone deposited by the action of rivers and streams or from glacial action. Sand and gravel deposited by rivers and streams are generally more satisfactory than glacial deposits because the former comprise rounded particles in a wide range of sizes and weaker materials have been eroded by the washing and abrasive action of moving water. Glacial deposits tend to have angular particles of a wide variety of sizes, poorly graded, which adversely affect the workability of a concrete in which they are used.

Crushed rock aggregates are generally more expensive than sand and gravel, owing to the cost of quarrying and crushing the stone. Providing the stone is hard, inert and well graded it serves as an admirable aggregate for concrete. The term 'granite aggregate' is used commercially to describe a wide range of crushed natural stones, some of which are not true igneous rocks. Natural granite is hard and dense and serves as an excellent aggregate.

Hard sandstone and close grained crystalline limestone when crushed and graded are commonly used as aggregate in areas where sand and gravel are not readily available.

Because of the depletion of inland deposits of sand and gravel, marine aggregates are used. They are obtained by dredging deposits of broken stone from the bed of the sea. Most of these deposits contain shells and salt. Though not normally harmful in reinforced concrete, limits should be set to the proportion of shells and salt in marine aggregates used for concrete. One of the disadvantages of marine fine aggregate is that it has a preponderance of one size of particle which can make design mix difficult. Sand from the beach is often of mainly single sized particles and contains an accumulation of salts. Beach sands to be used as fine aggregate in concrete should be carefully washed to reduce the concentration of salts.

Artificial aggregates

Blastfurnace slag is the by-product of the conversion of iron ore to pig iron and consists of the non-ferrous constituents of iron ore. The molten slag is tapped from the blastfurnace and is cooled and crushed. In areas where there is a plentiful supply of blastfurnace slag it is an economical and satisfactory aggregate for concrete.

Clean broken brick is used as an aggregate for concrete required to have a good resistance to damage by fire. The strength of the concrete produced with this aggregate depends on the strength and density of the bricks from which the aggregate is produced. Crushed engineering brick aggregate will produce a concrete of medium crushing strength. Porous brick aggregate should not be used for reinforced concrete work in exposed positions as the aggregate will absorb moisture and encourage the corrosion of the reinforcement.

Fine and coarse aggregate

Fine aggregate is the term used to describe natural sand, crushed rock and gravel, most of which passes through a 5 BS sieve and coarse aggregate the term used to describe natural gravel, crushed gravel or crushed rock, most of which is retained on a 5 BS sieve. The differentiation of fine and coarse aggregate is made because in practice the fine and coarse aggregate are ordered separately for mixing to produce a determined mix for particular uses and strengths of concrete.

Grading of aggregate

The word grading is used to describe the percentage of particles of a particular range of sizes in a given aggregate from fines (sand) to the largest particle size. A sound concrete is produced from a mix that can be readily placed and compacted in position, that is a mix that has good workability and after compaction is reasonably free of voids. This is affected by the grading of the aggregate and the water/cement ratio.

The grading of aggregate is usually given by the percentage by weight passing the various sieves used for grading. Continuously graded aggregate should contain particles graded in size from the largest to the smallest to produce a dense concrete. Sieve sizes from 75 to 5 (3 to 3/16 inches) are used for coarse aggregate.

An aggregate containing a large proportion of large particles is referred to as being 'coarsely' graded and one having a large proportion of small particles as 'finely' graded.

Particle shape and surface texture

The shape and surface texture of the particles of an aggregate affect the workability of a concrete mix. An aggregate with angular edges and a rough surface, such as crushed stone, requires more water in the mix to act as a lubricant to facilitate compaction than does one with rounded smooth faces to produce a concrete of the same workability. It is often necessary to increase the cement content of a mix made with crushed aggregate or irregularly shaped gravels to provide the optimum water/ cement ratio to produce concrete of the necessary strength. This additional water, on evaporation, tends to leave void spaces in the concrete which will be less dense than concrete made with rounded particle aggregate.

The nature of the surface of the particles of an aggregate will affect workability. Gravel dredged from a river will have smooth surfaced particles which will afford little frictional resistance to the arrangement of particles that takes place during compaction of concrete. A crushed granite aggregate will have coarse surfaced particles that will offer some resistance during compaction.

The shape of particles of aggregate is measured by an angularity index and the surface by a surface coefficient. Engineers use these to determine the true workability of a concrete mix which cannot be judged solely from the grading of particles.

Water

Water for concrete should be reasonably free from such impurities as suspended solids, organic matter and dissolved salts which may adversely affect the properties of concrete. Water that is fit for drinking is accepted as being satisfactory for mixing water for concrete.

CONCRETE MIXES

The strength and durability of concrete are affected by the voids in concrete caused by poor grading of aggregate, incomplete compaction or excessive water in the mix.

Water/cement ratio

Workability

The materials used in concrete are mixed with water for two reasons, firstly to enable the reaction with the cement which causes setting and hardening to take place and secondly to act as a lubricant to render the mix sufficiently plastic for placing and compaction.

About a quarter part by weight of water to one part by weight of cement is required for the completion of the setting and hardening process. This proportion of water to cement would result in a concrete mix far too stiff (dry) to be adequately placed and compacted. About a half by weight of water to one part by weight of cement is required to make a concrete mix workable.

It has been established that the greater the proportion of water to cement used in a concrete mix, the weaker will be the ultimate strength of the concrete. The principal reason for this is that the water, in excess of that required to complete the hardening of the cement, evaporates and leaves voids in the concrete which reduce its strength. It is practice, therefore, to define a ratio of water to cement in concrete mixes to achieve a dense concrete. The water/cement ratio is expressed as the ratio of water to cement by weight and the limits of this ratio for most concrete lie between 0.4 and 0.65. Outside these limits there is a great loss of workability below the lower figure and a loss of strength of concrete above the upper figure.

Water reducing admixtures

The addition of 0.2% by weight of calcium lignosulphonate, commonly known as 'lignin', to cement will reduce the amount of water required in concrete by 10% without loss of workability. This allows the cement content of a concrete mix to be reduced for a given water/cement ratio. Calcium lignosulphonate acts as a surface active additive that disperses the cement particles which then need less water to lubricate and disperse them in concrete.

Water reducing admixtures such as lignin are promoted by suppliers as densifiers, hardeners, water proofers and plasticisers on the basis that the reduction of water content leads to a more dense concrete due to there being fewer voids after the evaporation of water.

To ensure that the use of these admixtures does not adversely affect the durability of a concrete, it is practice to specify a minimum cement content.

Nominal mixes

Volume batching

The constituents of concrete may be measured by volume in batch boxes in which a nominal volume of aggregate and a nominal volume of cement are measured for a nominal mix, as for example in a mix of 1:2:4 of cement:fine:course aggregate. A batch box usually takes the form of an open top wooden box in which volumes of cement, fine and coarse aggregate are measured separately for the selected nominal volume mix. For a mix such as 1:2:4 one batch box will suffice, the mix proportions being gauged by the number of fillings of the box with each of the constituents of the mix.

Measuring the materials of concrete by volume is not an accurate way of proportioning and cannot be relied on to produce concrete with a uniformly high strength. Cement powder cannot be accurately proportioned by volume because while it may be poured into and fill a box, it can be readily compressed to occupy considerably less space. Proportioning aggregates by volume takes no account of the amount of water retained in the aggregate which may affect the water/cement ratio of the mix and affect the proportioning, because wet sand occupies a greater volume than does the same amount of sand when dry.

Volume batch mixing is mostly used for the concrete for the foundations and oversite concrete of small buildings such as houses. In these cases, the concrete is not required to suffer any large stresses and the strength and uniformity of the mix is relatively unimportant. The scale of the building operation does not justify more exact methods of batching.

Weight batching

A more accurate method of proportioning the materials of concrete is by weight batching, by proportioning the fine and coarse aggregate by weight by reference to the weight of a standard bag of cement. Where nominal mixes are weight batched it is best to take samples of the aggregate and dry them to ascertain the weight of water retained in the aggregate and so adjust the proportion of water added to the mix to allow for the water retained in the aggregate.

Water is incompressible and it is immaterial, therefore, whether it is proportioned by volume or by weight.

Designed mixes

Designed mixes of concrete are those where strength is the main criterion of the specified mix, which is judged on the basis of strength tests. The position in which concrete is to be placed, the means used and the ease of compacting it, the nature of the aggregate and the water/cement ratio all affect the ultimate strength of concrete.

A designed concrete mix is one where the variable factors are adjusted by the engineer to produce a concrete with the desired minimum compressive strength at the lowest possible cost. If, for example, the cheapest available local aggregate in a particular district will not produce a very workable mix it would be necessary to use a wet mix to facilitate placing and compaction, and this in turn would necessitate the use of a cement-rich mix to maintain a reasonable water/cement ratio. In this example it might be cheaper to import a different aggregate, more expensive than the local one, which would produce a comparatively dry but workable mix requiring less cement. These are the considerations the engineer and the contractor have in designing a concrete mix.

Prescribed mixes and standard mixes

Prescribed mixes and standard mixes are mixes of concrete where the constituents are of fixed proportion by weight to produce a 'grade' of concrete with minimum characteristics strength.

Mixing concrete

Concrete may be mixed by hand when the volume to be used does not warrant the use of mechanical mixing plant. The materials are measured out by volume in timber gauge boxes, turned over on a clean surface several times dry and then water is added as the mix is turned over several times until it has a suitable consistency and uniform colour. It is obviously difficult to produce mixes of uniform quality by hand mixing.

A small hand tilting mixer is often used. The mixing drum is rotated by a petrol or electric motor, the drum being tilted by hand to fill and empty it. This type of mixer takes over a deal of the back breaking work of mixing but does not control the quality of mixes as materials are measured by volume.

A concrete batch mixed mechanically feeds the materials into the drum where they are mixed and from which the wet concrete is poured. The materials may be either weight or volume batched.

For extensive works, plant is installed on site which stores cement delivered in bulk, measures the materials by weight and mechanically mixes them. Concrete for high strength reinforced concrete work can only be produced from batches (mixes) of uniform quality such as are produced by plant capable of accurately measuring and thoroughly mixing the materials.

Ready-mixed concrete is extensively used today. It is prepared in mechanical, concrete mixing depots where the materials are stored, weight batched and mixed and the wet concrete is transported to site on lorries on which rotating drums are mounted. The action of the rotating drum prevents the concrete from setting and hardening for an hour or more. Once delivered it must be placed and compacted quickly as it rapidly hardens. On cramped urban sites and where there is a nearby source of ready-mixed concrete this material is much used.

Placing and compacting concrete

The initial set of Portland cement takes place from half an hour to one hour after it is mixed with water. If a concrete mix is disturbed after the initial set has occurred the strength of the concrete may be adversely affected. It is usual to specify that concrete be placed as soon after mixing as possible and not more than half an hour after mixing.

A concrete mix consists of particles varying in size from powder to coarse aggregate graded to, say, 40. If a wet mix of concrete is poured from some height and allowed to fall freely the larger particles tend to separate from the smaller. This action is termed segregation of particles. Concrete should not, therefore, be tipped or poured into place from too great a height. It is usual to specify that concrete be placed from a height not greater than one metre.

Once in place concrete should be thoroughly consolidated or compacted. The purpose of compaction is to cause entrapped bubbles of air to rise to the surface in order to produce as dense and void-free concrete as possible. Compaction may be effected by agitating the mix with a spade or heavy iron bar. If the mix is dry and stiff this is a very laborious process and not very effective. A more satisfactory method is to employ a pneumatically operated poker vibrator which is inserted into the concrete and, by vibration, liberates air bubbles and compacts the concrete. As an alternative the formwork of reinforced concrete may be vibrated by means of a motor attached to it.

Construction joints

Because it is not possible, on most building sites, to place concrete continuously it is necessary to form construction joints. A construction joint is the junction of freshly placed concrete with concrete that has been previously placed, for example on the previous day. These construction joints are a potential source of weakness because there may not be a good bond between the two placings of concrete.

Practice is to brush the surface of concrete at construction joints soon after it has been placed to clean the surface to provide a bond to concrete subsequently placed. There should be as few construction joints as practical and joints should be either vertical or horizontal. Joints in columns are made as near as possible to beam haunching and those in beams at the centre or within the middle third of the span. Vertical joints are formed against a strip board.

Water bars are fixed across or cast into construction joints where there is a need to provide a barrier to the movement of water through the joint (see Chapter 1).

Curing concrete

Concrete gradually hardens and gains strength after its initial set. For this hardening process to proceed and the concrete to develop its maximum strength there must be water present in the mix. If, during the early days after the initial set, there is too rapid a loss of water the concrete will not develop its maximum strength. The process of preventing a rapid loss of water is termed curing concrete. Large exposed areas of concrete such as road surfaces are cured by covering the surface for at least a week after placing, with building paper, plastic sheets or wet sacks to retard evaporation of water. In very dry weather the surface of concrete may have to be sprayed with water in addition to covering it.

The formwork around reinforced concrete is often kept in position for some days after the concrete is placed in order to give support until the concrete has gained sufficient strength to be self-supporting. This formwork also serves to prevent too rapid a loss of water and so helps to cure the concrete. In very dry weather it may be necessary to spray the formwork to compensate for too rapid a loss of water.

Deformation of concrete

Hardened concrete will suffer deformation due to:

(1) Elastic deformation which occurs instantaneously and is dependent on applied stress
(2) Drying shrinkage that occurs over a long period and is independent of the stress in concrete
(3) Creep, which occurs over a long period and is dependent on stress in concrete
(4) Expansion and contraction due to changes in temperature and moisture
(5) ASR (alkali–silica reaction)

Elastic deformation

Under the stress of dead and applied loads of a building, hardened concrete deforms elastically. Vertical elements such as columns and walls are compressed and shorten in height and horizontal elements such as beams and floors lengthen due to bending. These comparatively small deformations which are related to the strength of the concrete are predictable and allowance is made in design.

Drying shrinkage

The drying shrinkage of concrete is affected principally by the amount of water in concrete at the time of mixing and to a lesser extent by the cement content of the concrete. It can also be affected by a porous aggregate losing water. Drying shrinkage is restrained by the amount of reinforcement in concrete.

The rate of shrinkage is affected by the humidity and temperature of the surrounding air, the rate of air flow over the surface and the proportion of surface area to volume of concrete.

Where concrete dries in the open air in summer, small masses of concrete will suffer about a half of the total drying shrinkage a month after placing and large masses about a half of the total shrinkage a year after placing. Shrinkage will not generally affect the strength or stability of a concrete structure, but is sufficient to require the need for movement joints where solid materials such as brick and block are built up to the concrete frame.

Creep

Under sustained load concrete deforms as a result of the mobility of absorbed water within the cement gel under the action of sustained stress. From the point of view of design, creep may be considered as an irrecoverable deformation that occurs with time at an ever decreasing rate under the action of sustained load. Creep deformation continues over very long periods of time to the extent that measurable deformation can occur thirty years after concrete has been placed. The factors that affect creep of concrete are the concrete mix, relative humidity and temperature, size of member and applied stress.

Concrete is a mix of aggregate, water and cement. Most aggregates used in dense concrete are inert and do not suffer creep deformation under load. The hardened cement water paste surrounding the particles of aggregate is subject to creep deformation under stress due to movements of absorbed water. The relative volume of cement gel to aggregate therefore affects deformation due to creep. Changing from a 1:1:2 to a 1:2:4 mix increases the volume of aggregate from 60 to 75% yet causes a reduction in creep by as much as 50%.

Temperature, relative humidity and the size of members have an effect on the hydration of cement and migration of water around the cement gel towards the surface of concrete. In general, creep is greater the lower the relative humidity and increases with a rise in temperature caused, for example, by solar heating. Small section members of concrete will lose water more rapidly than large members and will suffer greater creep deformation during the period of initial drying.

The effect of creep deformation has the most serious effect through stress loss in prestressed concrete, deflection increase in large span beams, buckling of slender columns and buckling of cladding in tall buildings.

ASR (alkali–silica reaction)

The chemical reaction of high silica-content aggregate with alkaline cement causes a gel to form, which expands and causes concrete to crack. The expansion, cracking and damage to concrete is often most severe where there is an external source of water in large quantities. Foundations, motorway bridges and concrete subject to heavy condensation have suffered severe damage through ASR.

The destructive effect of alkali–silica reaction has been known for some time. The damage caused by this reaction became apparent in this country with the report in 1980 of damage caused to a viaduct at Plymouth.

The expansion caused by the gel formed by the reaction is not uniform in time or location. The reaction may develop slowly in some structures yet very rapidly in others and may affect one part of a structure but not another. Changes in the method of manufacture of cement, that have produced a cement with higher alkalinity, are thought to be one of the causes of some noted failures. To minimise the effect of ASR it is recommended that cement rich mixes and high silica content aggregates be avoided.

REINFORCEMENT

In 1849 a French engineer, Joseph Monier, made some concrete flower tubs reinforced by casting wire mesh into the concrete and later, in 1867, he took out a patent for the process of strengthening concrete by embedding steel in concrete. Some years later François Hennebique applied Monier's idea to building and engineering when he developed reinforced concrete piles and later reinforced concrete structures. Today reinforced concrete is one of the two structural materials used in engineering and building works.

Concrete is strong in resisting compressive stress but comparatively weak in resisting tensile stress. The tensile strength of concrete is between one tenth and one twentieth of its compressive strength. Steel,

which has good tensile strength, is cast into reinforced concrete members in the position or positions where maximum tensile stress occurs.

To determine where tensile and compressive stresses occur in a structural member it is convenient to consider the behaviour of an elastic material such as India rubber under stress. A bar of rubber laid across, and not fixed to, two supports bends under load and the top surface shortens and becomes compressed under stress and the bottom surface stretched under tensile stress, as illustrated in Fig. 82. A member that is supported so that the supports do not restrain bending under load is said to be simply supported. From Fig. 82 it will be seen that maximum stretching due to tension occurs at the outwardly curved underside of the rubber bar. If the bar were of concrete it would seem logical to cast steel reinforcement in the underside of the bar. In that position the steel would be exposed to the surrounding air and it would rust and gradually lose strength. Further if a fire occurred in the building near the beam the steel might lose so much strength as to impair its reinforcing effect and the beam would collapse. It is practice, therefore, to cast the steel reinforcement into concrete so that there is at least 15 of concrete cover between the reinforcement and the surface of the concrete.

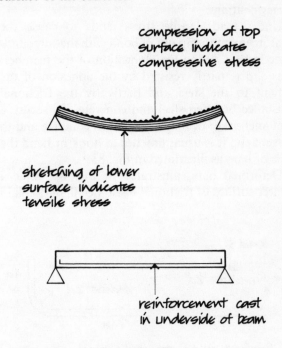

compression of top surface indicates compressive stress

stretching of lower surface indicates tensile stress

reinforcement cast in underside of beam

Simply supported beam

Fig. 82

Concrete cover

A 15 cover of concrete is sufficient to protect steel reinforcement from corrosion inside the majority of buildings and up to 60 where reinforced concrete is exposed to sea water and abrasion.

From laboratory tests and experience of damage caused by fires in buildings it has been established that various thicknesses of concrete cover will prevent an excessive loss of strength in steel reinforcement for particular periods of time. The presumption is that the concrete cover will protect the reinforcement for a period of time for the occupants to escape from the particular building during a fire. The statutory period of time that the concrete cover is to provide protection against damage by fire varies with the size and type of building from half an hour to four hours.

Bond and anchorage of reinforcement

The cement in concrete cast around steel reinforcement adheres to the steel just as it does to the particles of the aggregate and this adhesion plays its part in the transfer of tensile stress from the concrete to the steel. It is of importance, therefore, that the steel reinforcement be clean and free from scale, rust and oily or greasy coatings.

Under load, tensile stress tends to cause the reinforcement to slip out of bond with the surrounding concrete due to the elongation of the member. This slip is partly resisted by the adhesion of the cement to the steel and partly by the frictional resistance between steel and concrete. To secure a firm anchorage of reinforcement to concrete and to prevent slip it is usual practice to hook or bend the ends of bars as illustrated in Fig. 83.

Deformed bars, illustrated in Fig. 84, offer a greater surface of frictional resistance than do plain

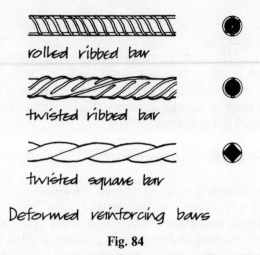

Deformed reinforcing bars

Fig. 84

bars and this can obviate the necessity to use hooked or bent ends for anchorage to prevent slip.

Shear

Beams are subjected to shear stresses due to the shearing action of the supports and the self-weight and imposed loads of beams. A pair of scissors does not cut paper, it shears it. The action of the blades, as they meet, is to force one side of the paper up and the other down and shear it into two pieces. The supports and the weight of the beam and its load act to shear a beam in the same way. Shear stress is greatest at the points of support and nil at mid-span in uniformly loaded beams. Shear failure occurs at an angle of 45° as illustrated in Fig. 85. Due to its poor tensile strength, concrete does not have great shear resistance and it is usual to introduce steel shear reinforcement in most beams of over, say, 2.5 span. The shear reinforcement may take the form of bars bent up at 45° near supports, or as steel stirrups or links more closely spaced at the point of support where maximum shear stress is developed, as illustrated in Fig. 85.

Fixed end support

A beam with fixed end support is restrained from simple bending by the fixed ends as illustrated in Fig. 35. Because of the upward, negative, bending close to the fixed ends the top of the beam is in tension while the underside is in tension at mid-span due to positive bending. In a concrete beam with fixed ends it is not sufficient to cast reinforcement in to the lower face of the beam only, as the concrete will not have sufficient tensile strength to resist tensile stresses in the top of

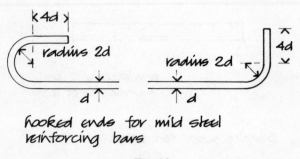

hooked ends for mild steel reinforcing bars

Fig. 83

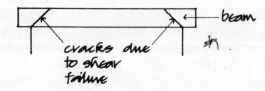

cracks due to shear failure

— beam

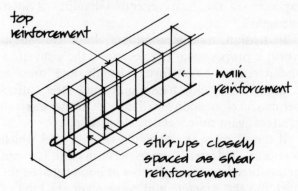

top reinforcement

main reinforcement

stirrups closely spaced as shear reinforcement

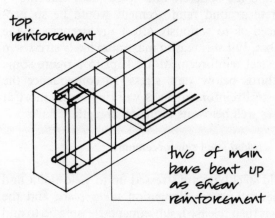

top reinforcement

two of main bars bent up as shear reinforcement

Shear reinforcement

Fig. 85

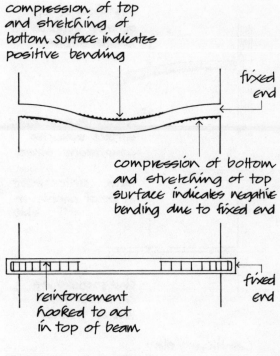

compression of top and stretching of bottom surface indicates positive bending

fixed end

compression of bottom and stretching of top surface indicates negative bending due to fixed end

reinforcement hooked to act in top of beam

fixed end

Beams with fixed ends

Fig. 86

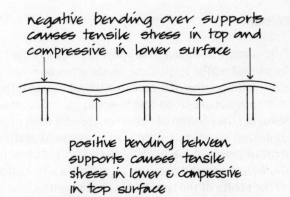

negative bending over supports causes tensile stress in top and compressive in lower surface

positive bending between supports causes tensile stress in lower & compressive in top surface

pairs of bars cranked up over support to act as top reinforcement

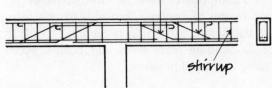

stirrup

Reinforced concrete beam to span continuously over supports

Fig. 87

the beam near points of support. Both top and bottom reinforcement are necessary, as illustrated in Fig. 86.

Similarly, a beam over several supports will bend as illustrated in Fig. 87, indicating normal or positive bending between supports and reverse or negative bending over supports. This indicates the stresses in a beam spanning continuously over supports, the reinforcement in such a beam being disposed as illustrated in Fig. 87.

Cantilever beams

A bar of rubber with one end fixed will bend under load as illustrated in Fig. 88, corrugation of the underside indicating compression and stretching of the top tensile stress. A comparable concrete canti-

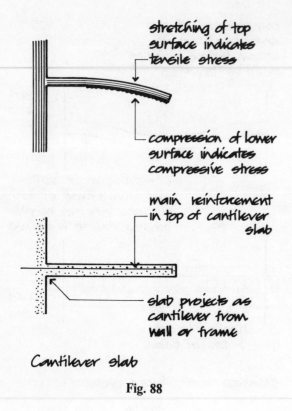

stretching of top
surface indicates
tensile stress

compression of lower
surface indicates
compressive stress

main reinforcement
in top of cantilever
slab

slab projects as
cantilever from
wall or frame

Cantilever slab

Fig. 88

lever will plainly require reinforcement in the top as illustrated in Fig. 88.

Columns

Columns are designed to support the loads of roofs, floors and walls. If all these loads acted concentrically on the section of the column then it would suffer only compressive stress and it would be sufficient to construct the column of either concrete by itself or of reinforced concrete to reduce the required section area. In practice, the loads of floor and roof beams and walls and wind pressure act eccentrically, that is off the centre of the section of columns and so cause some bending and tensile stress in columns. The steel reinforcement in columns is designed primarily to sustain compressive stress to reinforce the compressive strength of concrete, but also to reinforce the poor tensile strength of concrete against tensile stress due to bending from fixed end beams, eccentric loading and wind pressure.

Mild steel reinforcement

The cheapest and most commonly used reinforcement is round section mild steel rods of diameter from 6 to 40. These rods are manufactured in long lengths and can be quickly cut and easily bent without damage. The disadvantages of ordinary mild steel reinforcement are that if the steel is stressed up to its yield point it suffers permanent elongation, exposed to moisture it progressively corrodes and on exposure to the heat generated by fires it loses strength.

In tension, mild steel suffers elastic elongation which is proportional to stress up to the yield stress and it returns to its former length once stress is removed. At yield stress point mild steel suffers permanent elongation and then with further increase in stress again suffers elastic elongation.

If the permanent elongation of mild steel which occurs at yield stress were to occur in reinforcement in reinforced concrete, the loss of bond between the steel and the concrete and consequent cracking of concrete around reinforcement would be so pronounced as to seriously affect the strength of the member. For this reason maximum likely stresses in mild steel reinforcement are kept to a figure some two-thirds below yield stress. In consequence the mild steel reinforcement is working at the most at stresses well below its ultimate strength.

Cold worked steel reinforcement

If mild steel bars are stressed up to yield point and permanent plastic elongation takes place and the stress is then released, subsequent stressing up to and beyond the former yield stress will not cause a repetition of the initial permanent elongation at yield stress. This change of behaviour is said to be due to a reorientation of the steel crystals during the initial stress at yield point. In the design of reinforced concrete members, using this type of reinforcement, maximum stress need not be limited to a figure below yield stress, to avoid loss of bond between concrete and reinforcement, and the calculated design stresses may be considerably higher than with ordinary mild steel.

In practice it is convenient to simultaneously stress cold drawn steel bars up to yield point and to twist them axially to produce cold worked deformed bars with improved bond to concrete.

Deformed bars

To limit the cracks that may develop in reinforced concrete around mild steel bars, due to the stretching of the bars and some loss of bond under load, it is common to use deformed bars that have projecting ribs or are twisted to improve the bond to concrete.

The type of deformed reinforcing bars generally used are ribbed bars that are rolled from mild steel and ribbed along their length, ribbed mild steel bars that are cold drawn as high yield ribbed bars, ribbed, cold drawn and twisted bars, high tensile steel bars that are rolled with projecting ribs and cold twisted square bars. Figure 84 is an illustration of some typical deformed bars.

Galvanised steel reinforcing bars

Where reinforced concrete is exposed externally or is exposed to corrosive industrial atmospheres it is sound practice to use galvanised reinforcement as a protection against corrosion of the steel to prevent rust staining of fairface finishes and inhibit rusting of reinforcement that might weaken the structure. The steel reinforcing bars are cut to length, bent and then coated with zinc by the hot dip galvanising process. The considerable increase in cost of the reinforcement is well worth while.

Stainless steel reinforcement

Stainless steel is an alloy of iron, chromium and nickel on which an invisible corrosion resistant film forms on exposure to air. Stainless steel is about ten times the cost of ordinary mild steel. It is used for reinforcing bars in concrete where the cover of concrete for corrosion protection would be much greater than that required for fire protection and the least section of reinforced concrete is a critical consideration.

Assembling and fixing reinforcement

Reinforcing steel for concrete is used in the main to provide resistance to tensile stresses in structural members. The steel reinforcing bars must therefore be placed and secured in the positions inside formwork where they will be most effective in reinforcing concrete that will be poured and compacted inside the formwork around the reinforcement. It is of importance, therefore, that the reinforcement is rigidly fixed in position so that it is not displaced when wet concrete is placed and compacted.

Reinforcement for structural beams and columns is usually assembled in the form of a cage with the main and secondary reinforcement being fixed to

links or stirrups that hold it in position. The principal purpose of these links is to secure the longitudinal reinforcing bars in position when concrete is being placed and compacted. They also serve to some extent in anchoring reinforcement in concrete and in addition provide some resistance to shear with closely spaced links at points of support in beams.

Links are formed from small section reinforcing bars that are cut and bent to contain the longitudinal reinforcement. Stirrup or links are usually cold bent to contain top and bottom longitudinal reinforcement to beams and the main reinforcement to columns with the ends of each link overlapping, as illustrated in Fig. 89. As an alternative, links may be formed from two lengths of bar, the main 'U' shaped part of the link and a top section, as illustrated in Fig. 89. The advantage of this arrangement of links is that

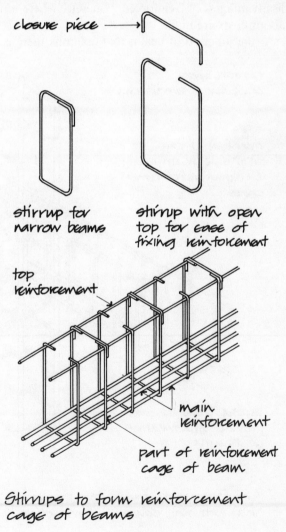

Stirrups to form reinforcement cage of beams

Fig. 89

where there are several longitudinal reinforcing bars in a cage they can be dropped in from the top of the links rather than being threaded through the links as the cage is wired up, thus saving time. Figure 89 is an illustration of part of a reinforcement cage for a reinforced concrete beam.

The separate cages of reinforcement for individual beams and column lengths are made up on site with the longitudinal reinforcement wired to the links with 1.6 mm soft iron binding wire that is cut to short lengths, bent in the form of a hair pin and looped and twisted around all intersections to secure reinforcing bars to links. The ends of binding wire must be flattened so that they do not protrude into the cover of concrete, where they might cause rust staining. Considerable skill, care and labour are required in accurately making up the reinforcing cages and assembling them in the formwork. This is one of the disadvantages of reinforced concrete where unit labour costs are high.

At the junction of beams and columns there is a considerable confusion of reinforcement, compounded by large bars to provide structural continuity at the points of support and cranked bars for shear resistance and as ties between members.

Figure 90 is an illustration of the junction of the reinforcement for a main beam with an external beam and an external column. It will be seen that the longitudinal bars for the beams finish just short of the column reinforcement for ease of positioning the beam cages and that continuity bars are fixed through the column and wired to beam reinforcement. The cranked 'U' bars fixed inside the column and wired to the main beam serve to anchor beam to column against lateral forces.

Figure 91 is an illustration of the reinforcement for the junction of four beams with a column. It will be seen that the reinforcement for intersecting beams is arranged to cross over at the intersection inside columns. Figure 91 is an illustration of a column splice made in vertical cages for convenience in erecting formwork floor by floor and handling cages.

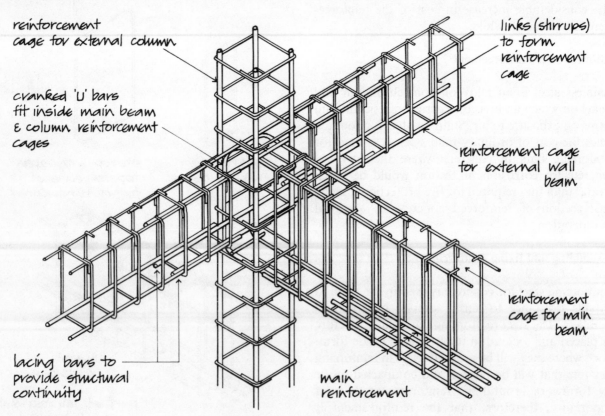

Reinforcement cages for reinforced concrete beam and external column connection

Fig. 90

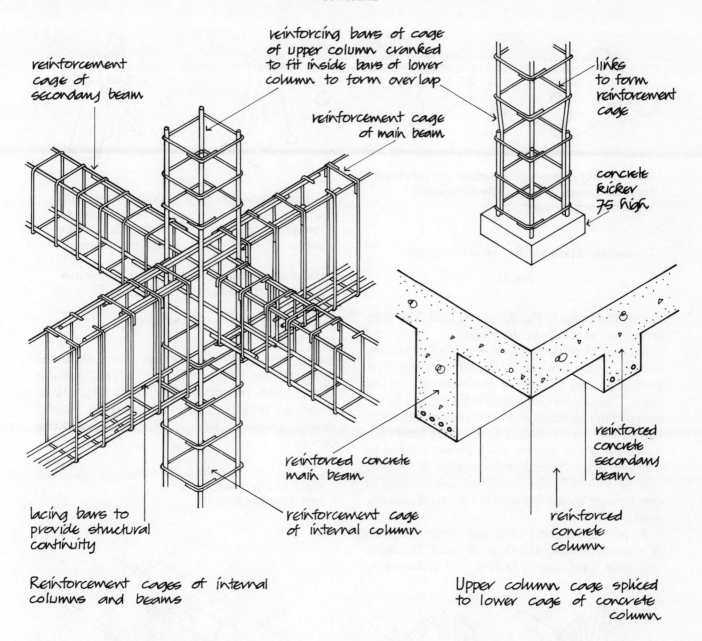

reinforcement cage of secondary beam

reinforcing bars of cage of upper column cranked to fit inside bars of lower column to form overlap

reinforcement cage of main beam

links to form reinforcement cage

concrete kicker 75 high

lacing bars to provide structural continuity

reinforced concrete main beam

reinforcement cage of internal column

reinforced concrete secondary beam

reinforced concrete column

Reinforcement cages of internal columns and beams

Upper column cage spliced to lower cage of concrete column

Reinforcement cages for reinforced concrete internal columns and beams

Fig. 91

In the reinforcement illustrated in Figs 90 and 91, the reinforcing bars are deformed to improve anchorage and obviate the necessity for hooked or bent ends of bars that considerably increase the labour of assembling reinforcing cages.

Spacers for reinforcement

To ensure that there is the correct cover of concrete around reinforcement to protect the steel from corrosion and to provide adequate fire protection, it is necessary to fix spacers to reinforcing bars between the bars and the formwork. These spacers must be securely fixed so that they are not displaced during placing and compacting of concrete and are strong enough to maintain the required cover of concrete. Concrete spacer blocks, the thickness of the required cover, can be cast on site from sand and cement with a loop of binding wire protruding for binding to reinforcement or one of the ready prepared concrete

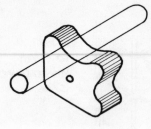
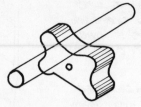

heavy duty spacer for horizontal or vertical reinforcement

spacer for vertical reinforcement

Concrete spacers for reinforcement

Fig. 92

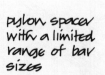

pylon spacer with a limited range of bar sizes

pylon spacer with a flexible grip to take a wide range of bar sizes

Plastic pylon (trestle chair) spacers for reinforcement

Fig. 94

spacers illustrated in Fig. 92 may be used. The holes in the spacers are for binding wire.

A range of ready prepared plastic wheel spacers and pylon spacers is available for fixing to reinforcement to provide a variety of thicknesses of concrete cover. These spacers, illustrated in Figs 93 and 94, are designed to clip firmly around various diameters of bar for both vertically and horizontally fixed reinforcement. These spacers, which are not affected by concrete, are sufficiently rigid to provide accurate spacing and will not cause surface staining of concrete, are commonly used in reinforced concrete work.

To provide support for top reinforcement in layers of concrete in slabs, steel chairs are used. The chairs are made from round section steel rods welded

together and galvanised. The chair spacer, illustrated in Fig. 95, sits on the lower layer of reinforcement and provides support for the upper layer. These chairs are robust enough to support the weights usually associated with placing and compacting concrete.

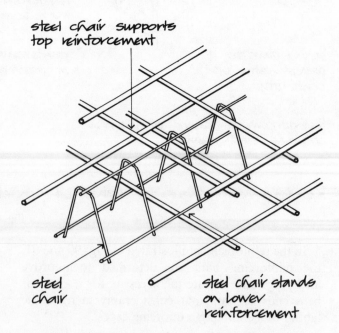

steel chair supports top reinforcement

steel chair

steel chair stands on lower reinforcement

Steel chair to provide spacing between layers of reinforcement

Fig. 95

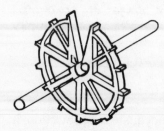

wheel spacer for vertical or horizontal reinforcement

wheel spacer for horizontal reinforcement

Plastic wheel spacers for reinforcement

Fig. 93

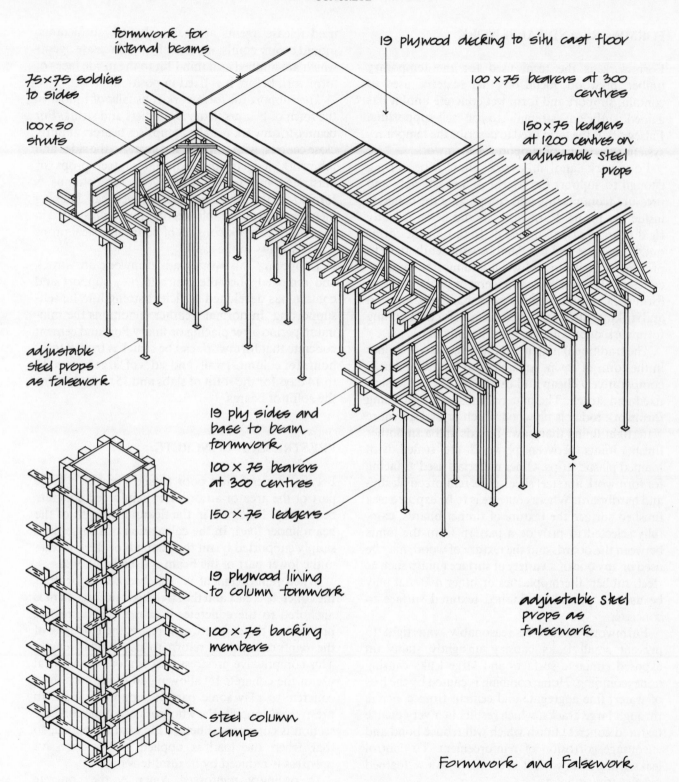

formwork for
internal beams

75×75 soldiers
to sides

100×50
struts

adjustable
steel props
as falsework

19 plywood decking to situ cast floor

100×75 bearers at 300
centres

150×75 ledgers
at 1200 centres on
adjustable steel
props

19 ply sides and
base to beam
formwork

100×75 bearers
at 300 centres

150×75 ledgers

19 plywood lining
to column formwork

100×75 backing
members

steel column
clamps

adjustable steel
props as
falsework

Formwork and Falsework

Fig. 96

FORMWORK AND FALSEWORK

Formwork is the term used for the temporary timber, plywood, metal or other material used to contain, support and form wet concrete until it has gained sufficient strength to be self-supporting. Falsework is the term used to describe the temporary system or systems of support for formwork.

Formwork and falsework should be strong enough to support the weight of wet concrete and pressure from placing and compacting the concrete inside the forms. Formwork should be sufficiently rigid to prevent any undue deflection of the forms out of true line and level and be sufficiently tight to prevent excessive loss of water and mortar from the concrete. The size and arrangement of the units of formwork should permit ease of handling, erection and striking. Striking is the term used for dismantling formwork once concrete is sufficiently hard.

The traditional material for formwork was timber in the form of sawn, square edged boarding that is comparatively cheap and can be readily cut to size, fixed and struck. The material most used for lining formwork today is plywood which provides a more watertight lining than sawn boards and a smoother finish. Joints between plywood are sealed with foamed plastic strips. Other materials used as facing for formwork are steel sheet, glass reinforced plastics and hardboard. Where concrete is to be exposed as a finished surface the texture of timber boards, carefully selected to provide a pattern from the joints between the boards and the texture of wood, may be used or any one of a variety of surface linings such as steel, rubber, thermoplastics or other material may be used to provide a finished textured surface to concrete.

Formwork should be reasonably watertight to prevent small leaks causing unsightly stains on exposed concrete surfaces and large leaks causing honeycombing. Honeycombing is caused by the loss of water, fine aggregate and cement from concrete through large cracks, which results in a very coarse textured concrete finish which will reduce bond and encourage corrosion of reinforcement. To control leaks from formwork it is common to use foamed plastic strips in joints.

To facilitate the removal of formwork and avoid damage to concrete as forms are struck, the surface of forms in contact with concrete should be coated with a release agent that prevents wet concrete adhering strongly to the forms. The more commonly used release agents are neat oils with surfactants, mould cream emulsions and chemical release agents which are applied as a thin film to the inside faces of formwork before it is fixed in position.

The support for formwork is usually of timber in the form of bearers, ledgers, soldiers and struts. For beams, formwork usually comprises bearers at fairly close centres, with soldiers and struts to the sides and falsework ledgers and adjustable steel props as illustrated in Fig. 96. Formwork for columns is formed with plywood facings, vertical backing members and adjustable steel clamps as illustrated in Fig. 96. Falsework consists of adjustable steel props fixed as struts to the sides.

Temporary falsework and formwork are struck and removed once the concrete they support and contain has developed sufficient strength to be self-supporting. In normal weather conditions the minimum period after placing ordinary Portland cement concrete that formwork can be struck is from 9 to 12 hours for columns, walls and sides of large beams, 11 to 14 days for the soffit of slabs and 15 to 21 days for the soffit of beams.

PRESTRESSED CONCRETE

Because concrete has poor tensile strength, a large part of the area of an ordinary reinforced concrete beam plays little part in the flexural strength of the beam under load. In the calculation of stresses in a simply supported beam the strength of the concrete in the lower part of the beam is usually ignored.

When reinforcement is stretched before or after the concrete is cast and the stretched reinforcement is anchored to the concrete, it causes a compressive prestress in the concrete as it resists the tendency of the reinforcement to return to its original length. This compressive prestress makes more economical use of the concrete by allowing all of the section of concrete to play some part in supporting load. In prestressed concrete the whole or part of the concrete section is compressed before the load is applied, so that when the load is applied the compressive prestress is reduced by flexural tension.

In ordinary reinforced concrete, the concrete around reinforcement is bonded to it and must, therefore, take some part in resisting tensile stress. Because the tensile strength of concrete is low it will crack around the reinforcement under load and when the load is removed the cracks will remain. The hair cracks on the surface of concrete are not only unsightly, they also reduce the protection against fire and corrosion the concrete cover is intended to give. In designing reinforced concrete members it is usual to limit the anticipated tensile stress in order to limit deflection and the extent of cracking of concrete around reinforcement. This is a serious limitation in the most efficient use of reinforced concrete, particularly in long span beams.

When reinforcement is stretched by prestressing and anchored to concrete and the prestress is released, the tendency of the reinforcement to return to its original length induces a compressive prestress in concrete. The stretching of reinforcement before it is cast into concrete is described as pre-tensioning and stretching after the concrete has been cast as post-tensioning. The advantage of the induced compressive prestress caused either by pre- or post-tensioning is that under load the tensile stress developed by bending is acting against the compressive stress induced in the concrete and in consequence cracking is minimised. If cracking of the concrete surface does occur and the load is reduced or removed, then the cracks close up due to the compressive prestress. Another advantage of the prestress is that the compressive strength of the whole of the section of concrete is utilised and the resistance to shear is considerably improved, so obviating the necessity for shear reinforcement.

Plainly, if the prestress is to be maintained the steel reinforcement must not suffer permanent elongation or creep under load as does mild steel. High tensile wire is used in prestressed concrete to maintain the prestress under load. Under load, a prestressed concrete member will bend or deflect and compressive and tensile stresses will be developed in opposite faces, as previously explained. Concrete in parts of the member will therefore have to resist compressive stress induced by the prestress as well as compressive stress developed during bending. For this reason high compressive strength concrete is used in prestressed work to gain the maximum advantage of the prestress. A consequence of the need to use high strength concrete is that prestressed members are generally smaller in section than comparable reinforced concrete ones.

Pre-tensioning

High tensile steel reinforcing wires are stretched between anchorages at each end of a casting bed and

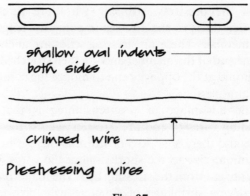

shallow oval indents both sides

crimped wire

Prestressing wires

Fig. 97

concrete is cast around the wires inside timber or steel moulds. The tension in the wires is maintained until the concrete around them has attained sufficient strength to take up the prestress caused by releasing the wires from the anchorages. The bond between the stretched wires and the concrete is maintained by the adhesion of the cement to the wires, by frictional resistance and the tendency of the wires to shorten on release and wedge into the concrete. To improve frictional resistance the wires may be crimped or indented, as illustrated in Fig. 97. When stressing wires are cut and released from the anchorages in the stressing frame the wires tend to shorten, and this shortening is accompanied by an increase in diameter of the wires which wedge into the ends of the member, as illustrated in Fig. 98.

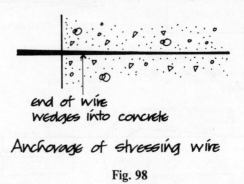

end of wire wedges into concrete

Anchorage of stressing wire

Fig. 98

Pre-tensioning of concrete is mainly confined to the manufacture of precast large span members such as floor beams, slabs and piles. The stressing beds required for this work are too bulky for use on site.

Post-tensioning

After the concrete has been cast inside moulds or formwork and has developed sufficient strength to resist the prestress, stressing wires are threaded through ducts or sheaths cast in along the length of the member. These prestressing wires are anchored at one end of the member and are then stretched and anchored at the opposite end to induce the compressive prestress.

The advantage of post-tensioning is that the stressing wires or rod are stressed against the concrete and there is no loss of stress as there is in pretensioning due to the shortening of the wires when they are cut from the stressing bed. The major part of the drying shrinkage of concrete will have taken place before it is post-tensioned and this minimises loss of stress due to shrinkage of concrete.

The systems of post-tensioning used are: Freyssinet, Gifford–Udall–CCL, Lee–McCall, Magnel–Blaton and the PSC one wire system.

The Freyssinet system

A number of high tensile steel wires, of diameter 7, are arranged around a core of fine coiled wire and threaded through a duct formed in the precast member. The duct is formed by casting an inflatable tube or greased rod into the concrete and withdrawing it when the concrete has set. The wires are held between a concrete cone cast into the concrete and a loose cone, as illustrated in Fig. 99. The wires are stressed by means of a jack and are then anchored by hammering the grooved cone into the cast-in cone and the wires are then released from the jack.

Cement and water grout is then forced under pressure into the cable duct to protect the wires from corrosion.

The Gifford–Udall–CCL system

High tensile wires, diameter 7, are threaded into a duct in the concrete member and are anchored to steel plates by means of barrels and wedges, as illustrated in Fig. 100. This system is designed to use from one to twelve wires, and each wire is separately

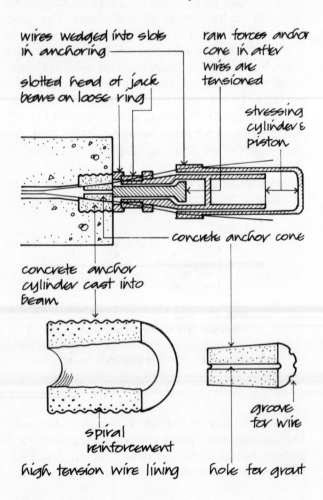

Post-tensioned prestressed concrete Freyssinet system

Fig. 99

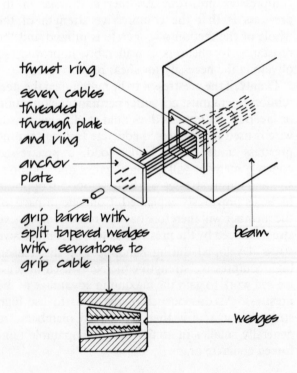

Post-tensioned pre-stressed concrete C.C.L. system

Fig. 100

stressed and anchored either at one or both ends of the member. When the wires have been stressed the duct through which they pass is filled with cement grout as before.

The advantage of this system is that the precise stress in each wire is controlled, whereas in the Freyssinet system all wires are jacked together and if one wire were to break the remaining wires would take up their share of the total stress and might be overstressed.

The Lee–McCall system

An alloy bar is threaded through a duct in the concrete member and stressed by locking a nut to one end and stressing the rod the other end with a jack and anchoring it with a nut. The simplicity of this system is self-evident.

The Magnel–Blaton system

High tensile wires are arranged in layers of four wires each and are held in position by metal spacers. The layers of wire are threaded through a duct in the concrete member. One end of the wires is fixed in metal sandwich plates against an anchor plate cast

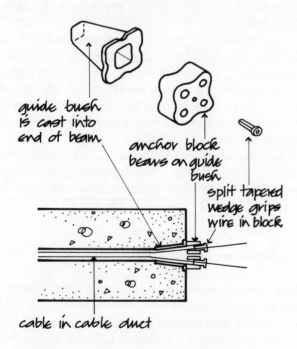

Post-tensioned pre-stressed concrete P. S. C. one wire system

Fig. 101

into the concrete. Pairs of wires are stressed in turn and wedged in position. The stressed wires are grouted in position in the duct by introducing cement grout through a hole in the top of the member leading to the duct.

The PSC one wire system

One, two or four high tensile wires are cast inside a sheath in the concrete member. The wires are stressed one at a time and anchored in taper sleeves which fit inside an anchor block, as illustrated in Fig. 101. After stressing the duct is grouted as previously described.

LIGHTWEIGHT CONCRETE

It is advantageous to employ lightweight concrete, such as no fines concrete, for the monolithic load-bearing walls of buildings and aerated concrete for structural members, such as roof slabs, supporting comparatively light loads, to combine the advantage of reduced deadweight and improved thermal insulation.

The various methods of producing lightweight concrete depend on:

(1) The presence of voids in the aggregate
(2) Air voids in the concrete
(3) Omitting fine aggregate or
(4) The formation of air voids by the addition of a foaming agent to the concrete mix

Lightweight aggregates containing voids were described in Chapter 6, Volume 2. The aggregates described for use in lightweight concrete building blocks are also used for mass concrete or reinforced concrete structural members, where improved thermal insulation is necessary and where the members, such as roof slabs, do not sustain large loads.

No fines concrete

No fines concrete consists of concrete made from a mix containing only coarse aggregate, cement and water. The coarse aggregate may be gravel, crushed brick or one of the lightweight aggregates. The coarse aggregate used in no fines concrete should be as near one size as practicable to produce a uniform distribution of voids throughout the concrete. To ensure a uniform coating of the aggregate particles with cement/water paste it is important that the aggregate be wetted before mixing and the maximum possible water/cement ratio, consistent with strength, be used to prevent separation of the aggregate and cement paste.

Construction joints should be as few as possible and vertical construction joints are to be avoided if practicable because successive placings of no fines concrete do not bond together firmly as do those of ordinary concrete.

Because of the porous nature of this concrete it must be rendered externally or covered with some protective coating or cladding material and the no fines concrete plastered or covered internally. A no fines concrete wall provides similar insulation to a sealed brick cavity wall of similar thickness. In Scandinavian countries no fines concrete walls, without reinforcement, have been used for multi-storey blocks of flats.

Aerated and foamed concretes

An addition of one part of powdered zinc or aluminium to every thousand parts of cement causes hydrogen to evolve when mixed with water. As the cement hardens a great number of small sealed voids form in the cement to produce aerated concrete, which usually consists of a mix of sand, cement and water.

Foamed concrete is produced by adding a foaming agent, such as resin soap, to the concrete mix. The foam is produced by mixing in a high speed mixer or by passing compressed air through the mix to encourage foaming. As the concrete hardens many sealed voids are entrained.

Aerated and foamed concretes are used for building blocks and lightweight roofing slabs, as described in Volume 3.

SURFACE FINISHES OF CONCRETE

It is only during the last fifty years that concrete has been accepted as a finish to buildings. Today a variety of concrete finishes is commonplace. The principal finishes employed are surfaces left untreated and either smooth or textured from the formwork, finishes textured by hammering and surfaces with an aggregate exposed.

Plain concrete finishes

Concrete is generally placed inside formwork in stages and when the formwork is removed variations in colour and texture and fine hair cracks usually clearly indicate the different placings of concrete. On drying, concrete shrinks and fine irregular shrinkage cracks appear in the surface in addition to the cracks and variations due to successive placings. One school of thought is to accept the cracks and variations in texture and colour as a fundamental of the material and make no attempt to control or mask them. Another school of thought is at pains to mask cracks and variations by means of designed joints and profiles on the surface.

Board marked concrete finishes are produced by compacting concrete by vibration against the surface of the timber formwork so that the finish is a mirror of the grain of the timber boards and the joints between them. This type of finish varies from the regular shallow profile of planed boards to the irregular marks of rough sawn boards and the deeper profile of boards that have been sand blasted to pronounce the grain of the wood. A necessary requirement of this type of finish is that the formwork be absolutely rigid to allow dense compaction of concrete to it and that the boards be non-absorbent.

One method of making construction joints is to form a horizontal indentation or protrusion in the surface of the concrete where construction joints occur by nailing a fillet of wood to the inside face of the timber forms or by making a groove in the boards so that the groove or protrusion in the concrete masks the construction joint.

Various plain concrete finishes can be produced by casting against plywood, hardboard or sheet metal to produce a flat finish or against corrugated sheets or crepe rubber to produce a profiled finish.

Tooled surface finishes

One way of masking construction joints, surface crazing of concrete and variations in colour is to tool the surface with hand or power operated tools. The action of tooling the surface is to break up the fine particles of cement and fine aggregate which find their way to the surface when wet concrete is compacted inside formwork and also to expose the coarse texture of aggregate.

Bush hammering

A round headed hammer with several hammer points on it is vibrated by a power driven tool which is held against the surface and moved successively over small areas of the surface of the concrete. The hammer crushes and breaks off the smooth cement film to expose a coarse surface. This coarse texture effectively masks the less obvious construction joints and shrinkage cracks.

Point tooling

A sharp pointed power vibrated tool is held on the surface and causes irregular indentations and at the same time spalls off the fine cement paste finish. By moving the tool over the surface a coarse pitted finish is obtained, the depth of pitting and the pattern of the pits being controlled by the pressure exerted and the movement of the tool over the surface. For best effect with this finish as large an aggregate size as possible should be used to maintain an adequate cover of concrete to reinforcement. The depth of the pitting should be allowed for in determining the cover required.

Dragged finish

A series of parallel furrows is tooled across the surface by means of a power operated chisel pointed tool. The depth and spacing of the furrows depend on the type of aggregate used in the concrete and the size of the member to be treated. This highly skilled operation should be performed by an experienced mason.

Margins to tooled finishes

Bush hammered and point tooled finishes should not extend to the edges or arrises of members as the hammering operation required would cause irregular and unsightly spalling at angles. A margin of at least 50 should be left untreated at all angles. As an alternative a dragged finish margin may be used with the furrows of the dragging at right angles to the angle.

Exposed aggregate finish

This type of finish is produced by exposing the aggregate of the concrete used in the member or by exposing a specially selected aggregate applied to the face or faces of the member. In order to expose the aggregate it is necessary either to wash or brush away the cement paste on the face of the concrete or to ensure that the cement paste does not find its way to the face of the aggregate to be exposed. Because of the difficulties of achieving this with in situ cast concrete, exposed aggregate finishes are confined in the main to precast concrete members and cladding panels.

One method of exposing the aggregate in concrete is to spray the surface with water, while the concrete is still green, to remove cement paste on the surface. The same effect can be achieved by brushing and washing the surface of green concrete. The pattern and disposition of the aggregate exposed this way is dictated by the proportioning of the mix and placing and compaction of the concrete, and the finish cannot be closely controlled.

To produce a distinct pattern or texture of exposed aggregate particles it is necessary to select and place the particles of aggregate in the bed of a mould or alternatively to press them into the surface of green concrete. This is carried out by precasting concrete.

Members cast face down are prepared by covering the bed of the mould with selected aggregate placed at random or in some pattern. Concrete is then carefully cast and compacted on top of the aggregate so as not to disturb the face aggregate in the bed of the mould. If the aggregate is to be exposed in some definite pattern it is necessary to bed it in water-soluble glue in the bed of the mould on sheets of brown paper that are washed off later. Once the concrete member has gained sufficient strength it is lifted from the mould and the face is washed to remove cement paste.

Large aggregate particles which are to be exposed are pressed into a bed of sand in the bed of the mould and the concrete is then cast on the large aggregate. When the member is removed from the mould after curing, the sand around the exposed aggregate is washed off. Alternatively, large particles may be pressed into the surface of green concrete and rolled, to bed them firmly and evenly.

CHAPTER FOUR

CONCRETE STRUCTURAL FRAMES

IN SITU CAST FRAMES

Joseph Aspdin produced the earliest Portland cement in order to manufacture artificial stone. A French gardener, Joseph Monier, who was making concrete flower boxes found that they cracked and to strengthen them he cast a wire mesh in the concrete. This was the birth of reinforced concrete and Monier took out a patent in 1867 for the manufacture of reinforced concrete flower pots. Some years later another Frenchman, François Hennebique, was chiefly responsible for the development of reinforced concrete for use in buildings, firstly as reinforced concrete piles and later as reinforced concrete beams and columns. In 1930, Freyssinet began development work that led to the use of prestressed concrete in building.

The first reinforced concrete framed building to be built in this country was the General Post Office building in London which was completed in 1910. Subsequently comparatively little use was made of reinforced concrete in this country until the end of the Second World War. Steel had been the traditional material used for structural frames and engineers regarded the newfangled reinforced concrete with some suspicion. The great shortage of steel that followed the end of the Second World War prompted engineers to use reinforced concrete as a substitute for steel in structural building frames.

The shortage of steel continued for some years after the end of the war. At the time the conventional method of providing fire protection to structural steel frames was to encase beams and columns in concrete that was cast in situ in formwork around the steel. This concrete casing added nothing to the strength of the steel members, added considerably to the dead weight of the frame and was costly in the formwork and falsework necessary for casting concrete.

With firstly a shortage and later the comparatively high cost of steel it was common to use a reinforced concrete structural frame with the concrete providing compressive strength and fire protection with the small section steel rods cast in to provide tensile strength where it was most needed. Up to the early 1980s the majority of framed buildings in this country were constructed with reinforced concrete frames. Recently a steel frame may be somewhat cheaper than a reinforced concrete frame for multistorey framed structures and wide span single storey shed buildings.

The reinforced concrete frames that were constructed after the Second World War were treated as a substitute for steel frames and were designed as though they were of steel, with columns on a rectangular grid supporting main floor beams supporting one way spanning floor slabs. This design procedure ignored the inherent differences between a steel and a concrete frame.

The conventional steel frame consists of 'I' section columns and beams that have greater strength on the axis parallel to the web of the section than at right angles to it. It is logical, therefore, to connect the main beams to the flanges of columns and span one way slabs between main beams with ties at right angles to main beams. The floor slabs bear on the beams in simple bending and do not act monolithically with the beams supporting them. In these conditions the rectangular column grid is the most economical arrangement.

The members of a reinforced concrete frame can be moulded to any required shape so that they can be designed to use concrete where compressive strength is required and steel reinforcement where tensile strength is required, and the members do not need to be of uniform section along their length or height.

The singular characteristics of concrete are that it is initially a wet plastic material that can be formed to any shape inside formwork for economy in section as a structural material, or for reasons of appearance, and when it is cast in situ will act monolithically as a rigid structure.

These characteristics are at once an advantage and a disadvantage. Unlimited choice of shape is an advantage structurally and aesthetically, but may well be a disadvantage economically in the complication of formwork and falsework necessary to form irregular shapes. A monolithically cast reinforced concrete frame has advantageous rigidity of connections in a frame and in a solid wall or shell structure, but this rigidity is a disadvantage in that it is less able to accommodate to movements due to settlement, wind pressure and temperature and moisture changes than is a more flexible structure.

The cost of formwork for concrete can be considerably reduced by repetitive casting in the same mould in the production of precast concrete cladding and structural frames, and the rigidity of the concrete frame can be of advantage on subsoils of poor or irregular bearing capacity and where severe earth movements occur as in areas subject to earthquakes.

The form of buildings, such as the shell forms illustrated in Volume 3, the Sydney Opera House, the Alvarado Palace Brazilia and precast cladding demonstrate the application of the initial plasticity of concrete and the structural rigidity and strength of reinforced concrete to advantage.

In spite of considerable publicity emphasising the advantages of steel as a structural frame material, the in situ cast reinforced concrete frame is still extensively used for both single and multi-storey buildings as a convenient and economic skeleton frame within which or on which a variety of wall envelopes may be supported or hung to provide the appearance of traditional loadbearing walling, panel cladding, infill framing and thin sheet finishes.

Structural frame construction

The principal use of reinforced in situ cast concrete as a structural material for building is as a skeleton frame of columns and beams with reinforced concrete floors and roof. In this use reinforced concrete differs little from structural steel skeleton frames cased in concrete. In those countries where unit labour costs are low and structural steel is comparatively expensive, a reinforced concrete frame is widely used as a frame for both single and multi-storey buildings such as the small framed building, with solid end walls and projecting balconies with upstands, illustrated in Fig. 102.

The in situ cast, reinforced concrete structural frame is much used for multi-storey buildings such as flats and offices. Repetitive floor plans can be formed inside a skeleton frame of continuous columns and floors. To use the same formwork and falsework, floor by floor, variations in the reinforcement and/or mix of concrete in columns, to support variations in loads, can provide a uniform column section. The

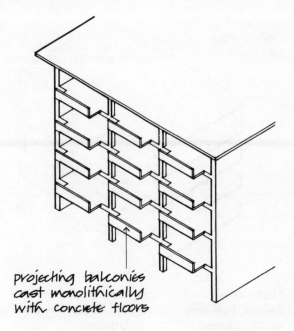

projecting balconies cast monolithically with concrete floors

In-situ cast reinforced concrete frame

Fig. 102

uniformity of column section and formwork makes for a speedily erected and economic structural frame.

An advantage of the reinforced concrete structural frame is that the columns, beams and floor slabs provide a level, solid surface on which walls and partitions can be built and between which walls, partitions and framing may be built and secured by bolting directly to a solid concrete backing.

A reinforced concrete structural frame with one way spanning floors is generally designed on a rectangular grid for economy in the use of materials in the same way as a structural steel frame. Where floors are cast monolithically with a reinforced concrete frame, the tie beams that are a necessary part of a structural steel frame may be omitted as the monolithically cast floors will act as ties. The flush floor slab soffit between main beams, illustrated in Fig. 103, may be of advantage in running services below ceilings and the disposition of demountable divisions.

The main beams in the external walls of a reinforced concrete frame may be cast as upstand beams, as illustrated in Fig. 103, which can serve as the apron below windows and the flush soffit below these beams will allow the head of windows to be in line with the ceiling and so obtain maximum penetration of light to the interior.

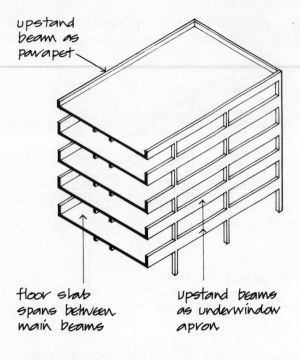

upstand
beam as
parapet

floor slab
spans between
main beams

upstand beams
as underwindow
apron

main beams
to columns to
centre access
corridor

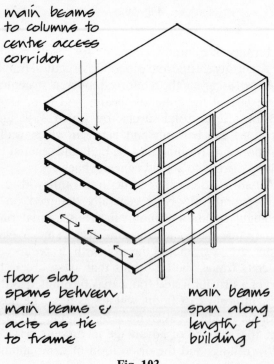

floor slab
spans between
main beams &
acts as tie
to frame

main beams
span along
length of
building

Fig. 103

Cross wall and box frame construction

Multi-storey structures, such as blocks of flats and hotels with identical compartments planned on successive floors one above the other, require per-

manent solid vertical divisions between compartments for privacy and sound and fire resistance.

In this type of building it is illogical to construct a frame and then build solid heavy walls within the frame to provide horizontal separation, with the walls taking no part in load bearing. A system of reinforced concrete cross walls at once provides sound and fire separation and acts as a structural frame supporting floors, as illustrated in Fig. 104. Between the internal cross walls reinforced concrete beam and slab or plate floors may be used. Where flats are planned on two floors as maisonettes the intermediate floor of the maisonette may be of timber joist and concrete beam construction to reduce cost and dead weight. The intermediate timber floor inside maisonettes is possible where building regulations require vertical and horizontal separation between adjacent maisonettes.

Where a system of cross walls and flat slab floors is employed the structure takes the form of a series of adjacent boxes, and this system is sometimes described as cross wall or box frame construction. A box frame form of construction may be used for external walls where the inherent strength and stability of a reinforced concrete wall are used, both for structural support and as an external wall perforated for

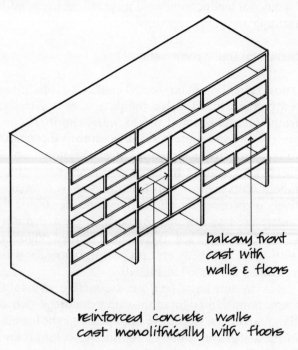

balcony front
cast with
walls & floors

reinforced concrete walls
cast monolithically with floors

Cross wall construction

Fig. 104

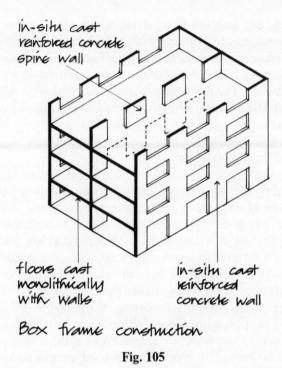

in-situ cast reinforced concrete spine wall

floors cast monolithically with walls

in-situ cast reinforced concrete wall

Box frame construction

Fig. 105

In addition to stiffening the whole building, such a service core will also carry a considerable part of floor loads by cantilevering floors from the core and using props in the form of slender columns on the face of the building. Similarly monolithically cast reinforced concrete flank end walls of slab blocks may be used to stiffen a skeleton frame structure against wind pressure on its long façade.

FLOOR CONSTRUCTION

In situ cast concrete floors

The principal types of reinforced in situ cast concrete floor construction are:

- Beam and slab
- Waffle grid slab
- Drop beam and slab
- Flat slab

Beam and slab floor

A beam and slab floor is generally the most economic and therefore most usual form of floor construction for reinforced concrete frames.

window openings as illustrated in Fig. 105. Here the wall frame may be used for external walls with an internal frame or as both external and internal walls.

Wind bracing

A steel frame depends on the use of continuously rolled, comparatively slender steel sections that are put together in the form of a skeleton frame. Where a part or parts of a building have to be enclosed, as for example lift shafts, stairs and lavatories, it is usual to construct the steel frame around these parts and then enclose them with brickwork carried at each floor by the frame. Initially concrete is a wet, plastic material that can be moulded to any desired shape and the shape is not dependent on the reinforcement which can be disposed to suit the shape of the concrete.

In a reinforced concrete structure it is not logical to cast beams and columns to support permanent brick walls when a monolithically cast concrete wall will serve the dual function of frame and enclosing wall. In multi-storey reinforced concrete framed buildings it is usual to contain the lifts, stairs and lavatories within a service core, contained in reinforced concrete walls, as part of the frame. The hollow reinforced concrete column containing the services and stairs is immensely stiff and will strengthen the attached skeleton frame against wind pressure.

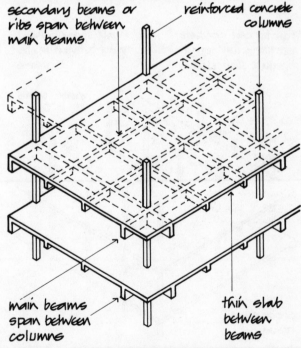

secondary beams or ribs span between main beams

reinforced concrete columns

main beams span between columns

thin slab between beams

Square grid beam and slab floor

Fig. 106

99

When a reinforced concrete frame is cast monolithically with reinforced concrete floors it is logical to design the floor slabs to span in both directions so that all the beams around a floor slab can bear part of the load. This two-way span of floor slabs effects some reduction in the overall depth of floors as compared to a one-way spanning floor slab construction. Since the most economical shape for a two-way spanning slab is square, the best column grid for a reinforced concrete frame with monolithically cast floors is a square one as illustrated in Fig. 106. The in situ cast reinforced concrete floor illustrated in Fig. 107 combines main and secondary beams as a grid to provide the least thickness of slab for economy in the mass of concrete used in construction. This square grid results in the minimum thickness of floor slab and minimum depth of beams and therefore the minimum dead weight of construction. Departure from the square column grid, because of user requirements and circulation needs in a building, will increase the overall depth, weight and therefore cost of construction of a reinforced concrete frame.

A rectangular column grid supports beams and a one-way span floor as illustrated in Fig. 107. The floor can be cast in situ on centering and falsework.

In a steel frame the skeleton of columns and beams is designed to carry the total weight and loads of the building, and the floor slabs, which span between beams, act independently of the frame. In in situ cast reinforced concrete frame and floor construction, columns, beams and floors are cast and act monolithically. The floor construction, therefore, acts with and affects the frame and should be considered as part of it.

Waffle grid slab floor

If the column grid is increased from about 6.0 to about 12.0 square or near square it becomes economical to use a floor with intermediate cross beams supporting thin floor slabs, as illustrated in Fig. 108. The intermediate cross beams are cast on a regular square grid that gives the underside of the floor the appearance of a waffle, hence the name. The advantage of the intermediate beams of the waffle is that they support a thin floor slab and so reduce the dead weight of the floor as compared to a flush slab of similar span. This type of floor is used where a widely spaced square column grid is necessary and floors support comparatively heavy loads. The economic span of floor slabs between intermediate beams lies between 900 and 3.5. The waffle grid form of the floor may be cast around plastic or metal formers laid on timber centering as illustrated in Fig. 108, so that the smooth finish of the soffit may be left exposed.

Drop slab floor

This floor construction consists of a floor slab which is thickened between columns in the form of a shallow but wide beam, as illustrated in Fig. 109. A drop slab floor is of about the same dead weight and cost as a comparable slab and beam floor and will have up to half the depth of floor construction from top of slab to soffit of beams. On a 12.0 square column grid the overall depth of a slab and beam floor would be about 1.2 where the depth of a drop slab floor would be about 600. This difference would cause a significant reduction in overall height of construction of a multi-storey building with appreciable saving in cost.

Flat slab (plate) floor

In this floor construction the slab is of uniform thickness throughout, without downstand beams and with the reinforcement more closely spaced

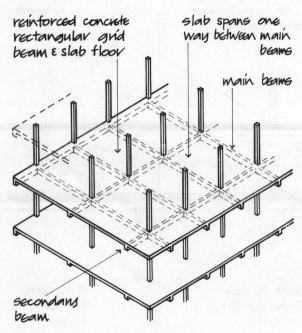

reinforced concrete rectangular grid beam & slab floor

slab spans one way between main beams

main beams

secondary beam

Rectangular grid beam and slab floor

Fig. 107

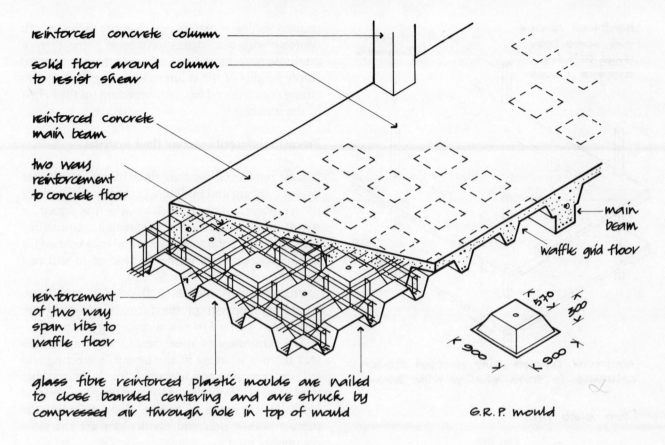

reinforced concrete column

solid floor around column
to resist shear

reinforced concrete
main beam

two way
reinforcement
to concrete floor

reinforcement
of two way
span ribs to
waffle floor

glass fibre reinforced plastic moulds are nailed
to close boarded centering and are struck by
compressed air through hole in top of mould

main
beam

waffle grid floor

570
300
900
900

G.R.P. mould

Waffle grid in-situ cast reinforced concrete floor

Fig. 108

between the points of support from columns. To provide sufficient resistance to shear at the junction of columns and floor, haunched or square headed columns are often formed, as illustrated in Fig. 110. The dead weight of this floor and its cost are greater than the floor systems previously described but its depth is less and this latter advantage provides the least overall depth of construction in multi-storey buildings.

The floor slabs in the floor systems described above may be of solid reinforced construction or constructed with one of the hollow, or beam or plank floor systems.

In modern buildings it is common to run air conditioning, heating, lighting and fire fighting services on the soffit of floors above a false ceiling and these services occupy some depth below which minimum floor heights have to be provided. Even though the beam and slab or waffle grid floors are the most economic forms of construction in themselves,

they may well not be the most advantageous where the services have to be fixed below and so increase the overall depth of the floor from the top of the slab to the soffit of the false ceiling below, because the services will have to be run below beams and so increase the depth between false ceiling and soffit of slab. Here it may be economic to bear the cost of a flat slab or drop slab floor in order to achieve the least overall height of construction and its attendant saving in cost.

Up to about a third of the cost of an in situ cast reinforced concrete frame goes to providing, erecting and striking the formwork and falsework for the frame and the centering for the floors. It is important, therefore, to maintain a uniform section of column up the height of the building and repetitive floor and beam design as far as possible, so that the same formwork may be used at each succeeding floor. Alteration of floor design and column section involves extravagant use of formwork. Uniformity of

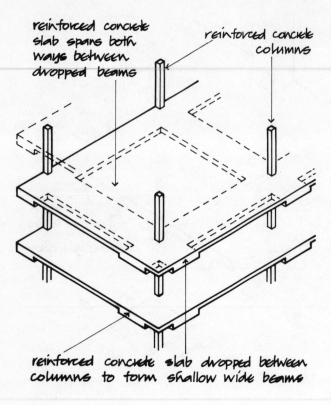

reinforced concrete slab spans both ways between dropped beams

reinforced concrete columns

reinforced concrete slab dropped between columns to form shallow wide beams

Drop slab floor

Fig. 109

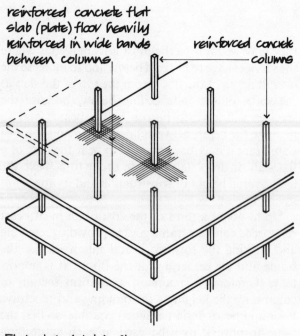

reinforced concrete flat slab (plate) floor heavily reinforced in wide bands between columns

reinforced concrete columns

Flat slab (plate) floor

Fig. 110

column section is maintained by using high strength concrete with a comparatively large percentage of reinforcement in the lower, more heavily loaded storey heights of the columns and progressively less strong concrete and less reinforcement up the height of the building.

Precast reinforced concrete floor systems

Precast reinforced concrete floor beams, planks, tee beams or beam and infill blocks that require little or no temporary support and on which a screed or structural concrete topping is spread are commonly used with structural steel frames and may be used for in situ cast concrete frames instead of in situ cast floors.

Precast beams and plank floors that require no temporary support in the form of centering are sometimes referred to as self-centering floors.

The advantage of these precast floor systems is that there is a saving in site labour in erecting and striking centering and falsework for floors and that there is no falsework in the form of props to obstruct work. Of the types of precast floor described the precast hollow unit and plank floor are the most commonly used.

Precast hollow floor units

These large precast reinforced concrete, hollow floor units are usually 400 or 1200 wide, 110, 150, 200, 250 or 300 thick and up to ten metres long for floors and thirteen and a half metres long for roofs. The purpose of the voids or hollows in the floor units is to reduce dead weight without affecting strength. The reinforcement is cast into the webs between hollows.

The hollow floor units can be used by themselves as floor slab with a floor screed or they may be used with a structural reinforced concrete topping with tie bars over beams for composite action with the beams. End bearing of these units is a minimum of 75 on steel and concrete beams and 100 on masonry and brick walls. Figure 111 is an illustration of precast hollow floor units.

Precast concrete plank floor units

These comparatively thin, prestressed solid plank, concrete floor units are designed as permanent shuttering and for composite action with structural reinforced concrete topping, as illustrated in Fig.

102

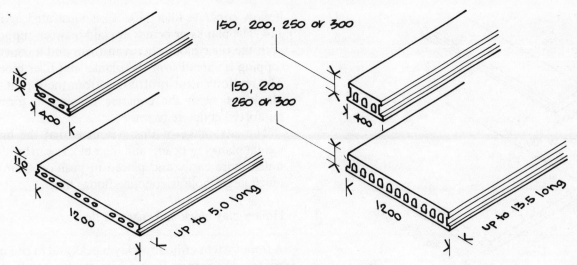

150, 200, 250 or 300

150, 200
250 or 300

400

1200

up to 5.0 long

400

1200

up to 13.5 long

Precast concrete floor units

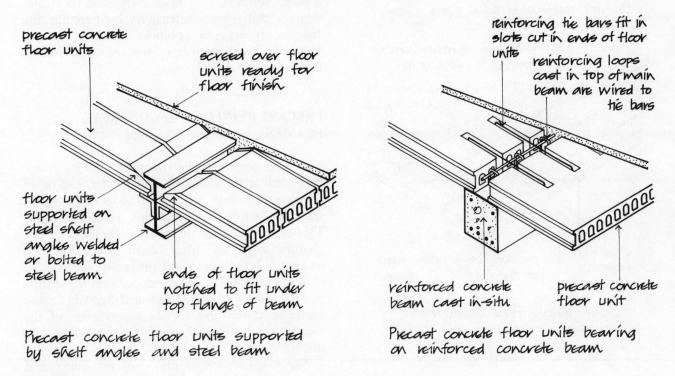

precast concrete
floor units

screed over floor
units ready for
floor finish

reinforcing tie bars fit in
slots cut in ends of floor
units

reinforcing loops
cast in top of main
beam are wired to
tie bars

floor units
supported on
steel shelf
angles welded
or bolted to
steel beam

ends of floor units
notched to fit under
top flange of beam

reinforced concrete
beam cast in-situ

precast concrete
floor unit

Precast concrete floor units supported
by shelf angles and steel beam

Precast concrete floor units bearing
on reinforced concrete beam

Hollow precast reinforced concrete floor units

Fig. 111

112. The units are 400 or 1200 wide, 65, 75 or 100 thick and up to 9½ metres long for floors and 10 metres for roofs. It may be necessary to provide some temporary propping to the underside of these planks until the concrete topping has gained sufficient strength.

Precast concrete tee beams

Precast prestressed concrete tee beam floors are mostly used for long span floors in such buildings as stores, supermarkets, swimming pools and multi-storey car parks where there is a need for wide span

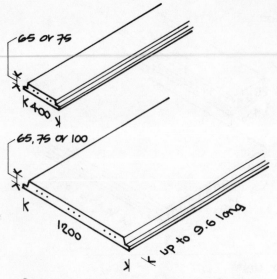

Precast prestressed solid plank floor units

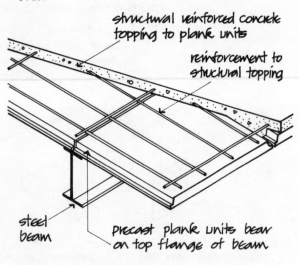

structural reinforced concrete topping to plank units

reinforcement to structural topping

steel beam

precast plank units bear on top flange of beam

Precast plank floor units for composite construction floor

Fig. 112

floors and the depth of this type of floor is not a disadvantage. The floor units are cast in the form of a double tee, as illustrated in Fig. 113. The strength of these units is in the depth of the ribs which support and act with the comparatively thin top web. A structural reinforced concrete topping is cast on top of the floor units.

Precast beam and filler block floor

This floor system consists of precast reinforced concrete planks or beams that support precast

hollow concrete filler blocks, as illustrated in Fig. 114. The planks or beams are laid between supports with the filler blocks between them and a concrete topping is spread over the planks and filler blocks. The reinforcement protruding from the top of the plank acts with the concrete topping to form a reinforced concrete beam.

The advantage of this system is that the lightweight planks or beams and filler blocks can be lifted much more easily and placed in position than the much larger hollow concrete floor units.

Hollow clay block and concrete floor

A floor system of hollow clay blocks and in situ cast reinforced concrete beams between the blocks and concrete topping, cast on centering and falsework, was for many years extensively used for the fire resisting properties of the blocks. This floor system is much less used because of the very considerable labour in laying the floor.

PRECAST REINFORCED CONCRETE FRAMES

The development of the structural building frame depended on the use of steel and, later, reinforced concrete as a skeleton frame to support floors and walls of the traditional materials, stone and brick. This form of construction was facilitated by a plentiful supply of cheap labour to construct the frame and build walls using the traditional skills of masonry and bricklaying developed over the centuries. The supply of skilled and unskilled labour was adequate at the time for the construction of the comparatively few large buildings in cities and towns.

The extensive programmes of rebuilding and rehousing that followed the end of the Second World War coincided with a shortage of the traditional building materials, i.e. brick, stone, timber and steel, and a depleted labour force wholly inadequate to the scale of the projected work. In the event, a substantial part of the building programmes was met by the use of concrete in the form of precast frames, cladding units and wall frames. The combination of the use of precast concrete units and the introduction of the tower crane made it possible to produce standard components and assemble them on site with the minimum of skilled labour.

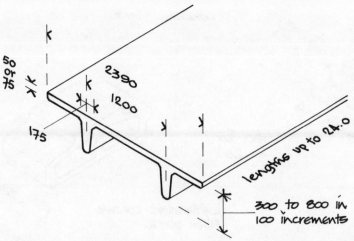

Precast prestressed reinforced concrete double tee beam

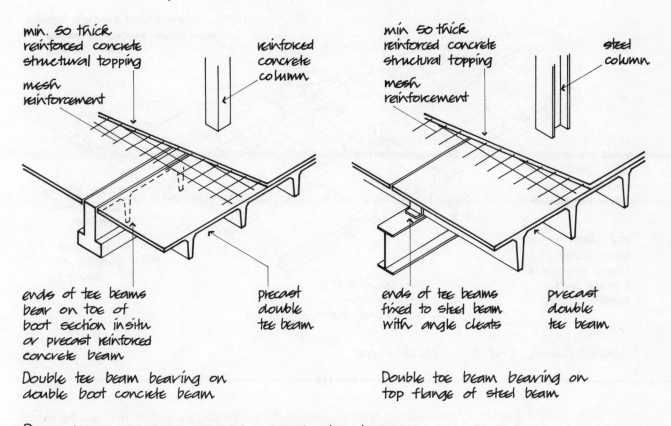

min. 50 thick reinforced concrete structural topping

mesh reinforcement

reinforced concrete column

ends of tee beams bear on toe of boot section in situ or precast reinforced concrete beam

precast double tee beam

Double tee beam bearing on double boot concrete beam

min. 50 thick reinforced concrete structural topping

mesh reinforcement

steel column

ends of tee beams fixed to steel beam with angle cleats

precast double tee beam

Double tee beam bearing on top flange of steel beam

Precast prestressed concrete double tee beam

Fig. 113

The advantage of precast concrete was that the materials cement, sand, gravel and crushed rock were readily available and could be combined with the least amount of steel to produce a material that was structurally sound and could serve both as a frame and as a wall material for buildings. Mechan-

isation of the production of standard, repetitively cast units off the site and their assembly on site required the least amount of labour, either skilled or unskilled.

The two basic forms of precast concrete frames are (a) the frame members precast either as separate

105

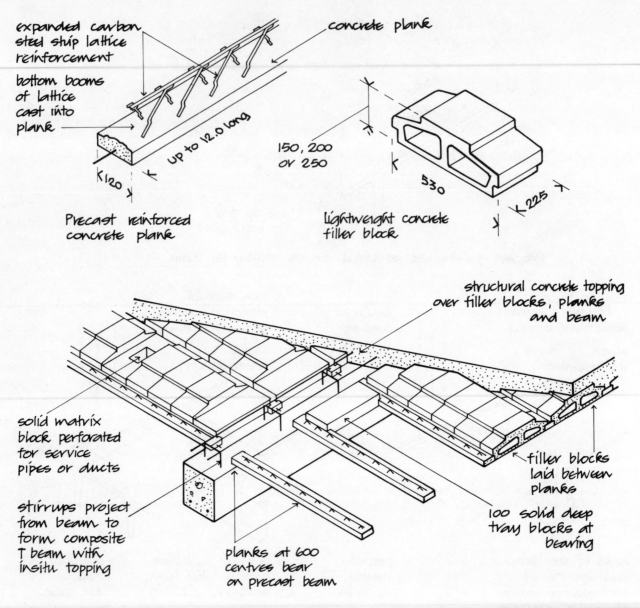

expanded carbon steel strip lattice reinforcement

concrete plank

bottom booms of lattice cast into plank

up to 12.0 long

Precast reinforced concrete plank

K 120

150, 200 or 250

530

K 225

Lightweight concrete filler block

structural concrete topping over filler blocks, planks and beam

solid matrix block perforated for service pipes or ducts

stirrups project from beam to form composite T beam with insitu topping

planks at 600 centres bear on precast beam

filler blocks laid between planks

100 solid deep tray blocks at bearing

Precast beam and filler block floor

Fig. 114

lengths of column and beam or combined as beam and column units, which are assembled and joined on site ready for infill panels or a cladding of brick or precast concrete units and (b) the wall frame units which serve both as structure and walling combined in large wall units, usually storey height, assembled and joined on site to precast concrete floor units.

The advantage of the precast frame is that it allows the use of various wall and cladding finishes, either as infill panels fixed inside the frame or as a covering of cladding. The wall frame, however, is a large, solid,

storey height panel of concrete that can be varied only in the surface finish of concrete and the size and disposition of windows.

In the years from 1950 to 1970 precast concrete frames, wall frames and cladding were extensively used as an accepted form of construction for multi-storey housing and other large buildings where standard units of construction could be used to advantage. Since the early 1970s, precast concrete has lost favour as a material for framing and cladding buildings. This loss of favour is principally a change of fashion as glass reinforced cement, then plastics and more recently sheet metal cladding panels were used instead of precast concrete, and for reasons of economy steel was used for structural frames in place of concrete. Publicised failures of precast concrete wall units, such as that at Ronan Point, have done much to add to the disfavour of concrete as a building material. These failures were not failures of the material itself but failures of the systems of jointing due to poor design or faulty workmanship or a combination of both.

Precast concrete has been established as a sound, durable material for framing and cladding buildings where repetitive casting of units is an acceptable and economic form of construction. Fashions change and it is likely that precast concrete will once again find favour as a building material.

The chief problem in precast concrete framework is joining the members on site, particularly if the frame is to be exposed, to provide a solid, rigid bearing in column joints and a strong, rigid bearing of beams to columns that adequately ties beams to columns for structural rigidity.

Where the frame is made up of separate precast column and beam units there is a proliferation of joints to be made on site. The number of site joints is reduced by the use of precast units that combine two or more column lengths with beams, as illustrated in Fig. 115. The number of columns and beams that can be combined in one precast unit depends on the particular design of the building and the facilities for casting, transporting, hoisting, and fixing units on site.

The general arrangement of precast structural units is as separate columns, often two storey height and as cruciform, 'H' or 'M' frames. The 'H' frame unit is often combined with under window walling, as illustrated in Fig. 116.

The two basic systems of jointing used for connections of column to column are by direct end bearing or by connection to a bearing plate welded to protruding studs. Direct bearing of ends is effected through a locating dowel which can also be used as a post tensioning connection, as illustrated in Fig. 117. A coupling plate connection is made by welding a plate to studs protruding from the end of one column and bolting studs protruding from the other to the plate, as illustrated in Fig. 116. Plainly the studs and plate must be accurately located else there will be an excessive amount of site labour in making this connection. The completed joint is usually finished by casting concrete around the joint. Alternatively the joint may be made with bronze studs and plate and left exposed as a feature of an externally exposed frame.

Beams are joined to columns by bearing on a haunch cast on the column or by connecting a steel box or plate, cast in the ends of beams at an angle or plate set in a housing in columns, as illustrated in Fig. 118.

Precast floor units bear either directly on beams or, more usually, on supporting nibs cast for the purpose. Ends of floor units are tied to beams through protruding studs or dowels so that the floor units serve to transfer wind pressure back to an in situ cast service and access core.

The precast frame illustrated in Fig. 115 is the structural external wall system for a 22 storey block of flats. The precast structural framework was built around a central in situ cast reinforced concrete access and service core containing lifts and stairs. The precast framework is tied to the central core through the precast concrete floor units at each floor level which are dowel fixed to the precast frame and tied with reinforcement to the in situ core. The precast framework is vertically tensioned by couplers through columns, as illustrated in Fig. 117, so that column ends are compressed to the dry mortar bed. The storey height frames are linked by short lengths of beam that are dropped in and tied to frames.

The precast framework was designed for rapid assembly through precasting and direct bearing of beams on columns and end bearing of columns, to avoid the use of in situ cast joints that are laborious to make and which necessitate support of beams while the in situ concrete hardens.

The top hung, exposed aggregate, precast concrete cladding panels have deep rebate horizontal joints and open vertical joints with mastic seals to columns, as illustrated in Fig. 119.

The precast reinforced concrete frame illustrated in Fig. 118 is a proprietary frame system for use in framing low rise slab blocks where the frame serves as the structural walls that are tied through the hollow precast concrete floor units that are tied to the frame. These standard frame sections can be used for various systems of solid and panel wall cladding systems.

Precast concrete wall frames

Precast concrete wall frames were used extensively in

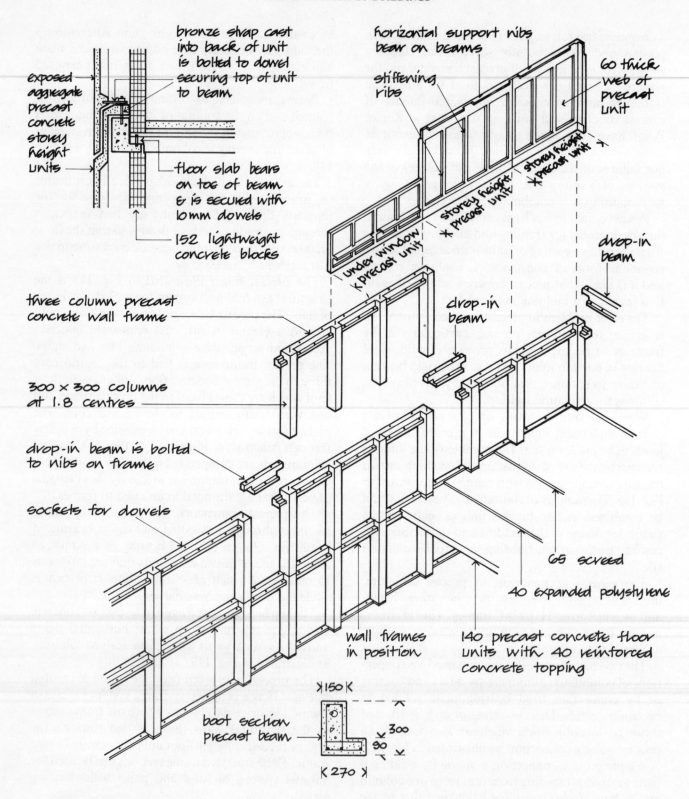

exposed aggregate precast concrete storey height units

bronze strap cast into back of unit is bolted to dowel securing top of unit to beam

horizontal support ribs bear on beams

stiffening ribs

60 thick web of precast unit

floor slab bears on toe of beam & is secured with 10mm dowels

152 lightweight concrete blocks

under window precast unit

storey height precast unit

storey height precast unit

drop-in beam

three column precast concrete wall frame

drop-in beam

300 × 300 columns at 1.8 centres

drop-in beam is bolted to nibs on frame

sockets for dowels

wall frames in position

65 screed

40 expanded polystyrene

140 precast concrete floor units with 40 reinforced concrete topping

boot section precast beam

150

300

90

270

Precast reinforced concrete wall frames and precast concrete cladding units

Fig. 115

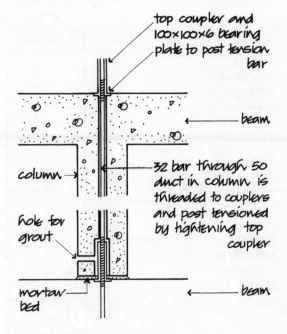

Precast concrete frame to frame joint

Fig. 117

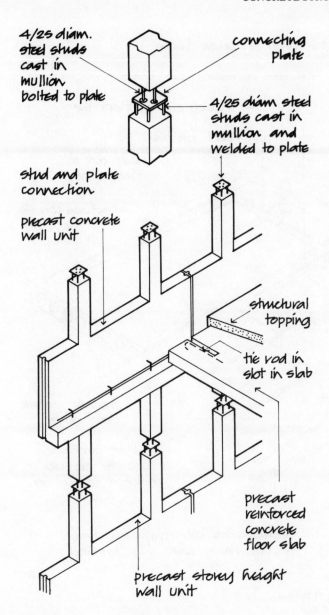

Precast concrete wall units

Fig. 116

Russia and northern European countries in the construction of multi-storey housing where repetitive units of accommodation were framed and enclosed by large precast reinforced concrete wall panels that served as both external and internal walls and as a structural frame. The advantages of this system of building are that large, standard, precast concrete wall units can be cast off site and rapidly assembled on site largely independent of weather conditions, a prime consideration in countries where temperatures are below freezing for many months of the year.

Reinforced concrete wall frames can support the loads of a multi-storey building, can be given an external finish of exposed aggregate or textured finish that requires no maintenance, can incorporate insulation either as a sandwich or lining and have an internal finish ready for decoration. Window and door openings are incorporated in the panels which can be delivered to site with windows and doors fixed in position.

In this system of construction the prime consideration is the mass production of complete wall units off the site, under cover, by unskilled or semi-skilled labour assisted by mechanisation as far as practical towards the most efficient and speedy erection of a building. The appearance of the building is a consequence of the chosen system of production and erection rather than a prime consideration.

The concrete wall units will give adequate protection against wind and rain by the use of rebated horizontal joints and open drained vertical joints with back-up air seals similar to the joints used with precast concrete cladding panels.

Some systems of wall frame incorporate a sandwich of insulation in the thickness of the panel with the two skins of concrete tied together across the insulation with non-ferrous ties. This is not a very satisfactory method of providing insulation as a

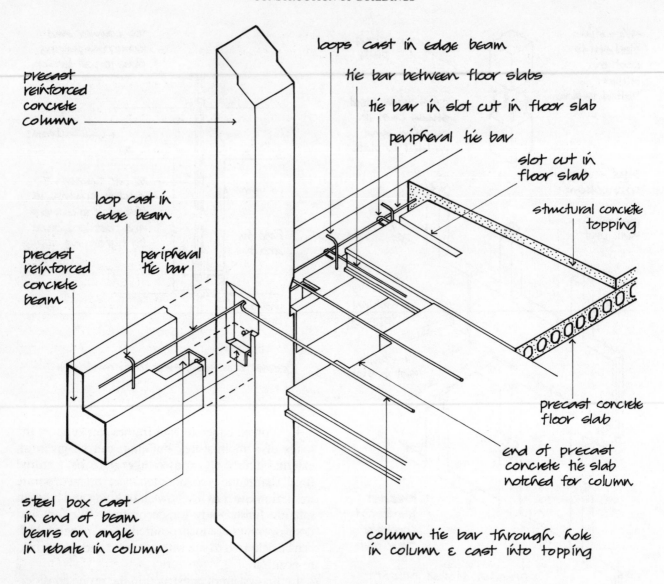

precast
reinforced
concrete
column

loops cast in edge beam

tie bar between floor slabs

tie bar in slot cut in floor slab

peripheral tie bar

slot cut in
floor slab

structural concrete
topping

loop cast in
edge beam

precast
reinforced
concrete
beam

peripheral
tie bar

precast concrete
floor slab

end of precast
concrete tie slab
notched for column

steel box cast
in end of beam
bears on angle
in rebate in column

column tie bar through hole
in column & cast into topping

Precast reinforced concrete structural frame

Fig. 118

sufficient thickness of insulation for present day standards will require substantial ties between the two concrete skins and the insulation may well absorb water from drying out of concrete and rain penetration, and so be less effective as an insulant. For best effect the insulation should be applied to the inside face of the wall as an inner lining to panels, or as a site fixed or built inner lining or skin.

The wall frame system of construction depends, for the structural stability of the building, on the solid, secure bearing of frames on each other and the firm bearing and anchorage of floor units to the wall

frames and back to some rigid component of the structure, such as in situ cast service and access cores.

Figure 120 is an illustration of a typical precast concrete wall frame.

LIFT SLAB CONSTRUCTION

In this system of construction the flat roof and floor slabs are cast one on the other at ground level around columns or in situ cast service, stair and lift cores. Jacks operating from the columns or cores pull the roof and floor slabs up into position.

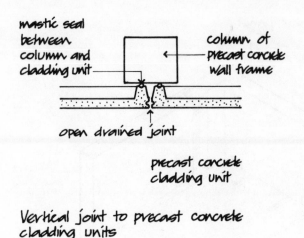

mastic seal between column and cladding unit

column of precast concrete wall frame

open drained joint

precast concrete cladding unit

Vertical joint to precast concrete cladding units

Fig. 119

This system of construction was first employed in America in 1950. Since then many buildings in America, Europe and Australia have been constructed by this method. The advantage of the system is that the only formwork required is to the edges of the slabs and no centering whatever is required to the soffit of roof or floors. The slabs are cast monolithically and can be designed to span continuously between and across points of support, and so employ the least thickness of slab. Where it is convenient to cantilever slabs beyond the edge columns and where cantilevers for balconies, for example, are required they can, without difficulty, be arranged as part of the slab.

The advantages of this system are employed most fully in simple, isolated point block buildings of up to five storeys where the floor plans are the same throughout the height of the building and a flush slab floor may be an advantage. The system can be employed for beam and slab and waffle grid floors, but the forms necessary between the floors to give the required soffit take most of the advantage of simplicity of casting on the ground.

Steel or concrete columns are first fixed in position and rigidly connected to the foundation and the ground floor slab is then cast. When it has matured it is sprayed with two or three coats of a separating medium consisting of wax dissolved in a volatile spirit. As an alternative, polythene sheet or building paper may be used as a separating medium. The first floor slab is cast inside edge formwork on top of the ground floor slab and when it is mature it is in turn coated or covered with the separating medium and the next floor slab is cast on top of it. The casting of successive slabs continues until all the floors and roof

have been cast one on the other on the ground. Lifting collars are cast into each slab around each column.

The slabs are lifted by jacks, operating on the top of each column, which lift a pair of steel rods attached to each lifting collar in the slab being raised. A central control synchronises the operation of the hydraulically operated reciprocating ram type jacks to ensure a uniform and regular lift.

The sequence of lifting the slabs depends on the height of the building, the weight of the slabs and extension columns, the lifting capacity of the jacks and the cross-sectional area of the columns during the initial lifting. The bases of the columns are rigidly fixed to the foundations so that when lifting commences the columns act as vertical cantilevers. The load that the columns can safely support at the beginning of the lift limits the length of the lower column height and the number of slabs that can be raised at one time. As the slabs are raised they serve as horizontal props to the vertical cantilever of the columns and so increasingly stiffen the columns, the higher the slabs are raised. The sequence of lifting illustrated in Fig. 121 is adopted so that the roof slab, which is raised first, stiffens the columns which are then capable of taking the load of the two slabs subsequently lifted, as illustrated. The lifting sequence for a three-storey building is illustrated in Fig. 121 where the roof is raised on the lower column lengths, followed by the upper floor slabs. The steel lifting collars which are cast into each slab around each column provide a means of lifting the slabs and also act as shear reinforcement to the slabs around columns, and so may obviate the necessity for shear reinforcement to the slabs. Figure 122 is an illustration of a typical lifting collar fabricated from mild steel angle sections welded together and stiffened with plates welded in the angle of the sections.

The lifting collars are fixed to steel columns by welding shear blocks to plates welded between column flanges and to the collar after the slab has been raised into position, as illustrated in Fig. 123. Connections to concrete columns are made by welding shear blocks to the ends of steel channels cast into the column and by welding the collar to the wedges, as illustrated in Fig. 123. With this connection it is necessary to cast concrete around the exposed steel wedges for fire protection.

The connection of steel extension columns is made by welding, bolting or riveting splice plates to the flanges of columns at their junction. Concrete exten-

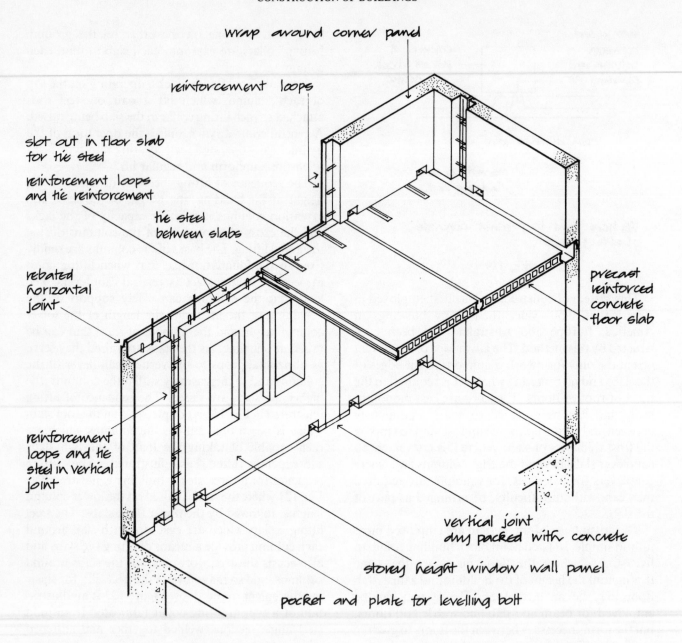

wrap around corner panel

reinforcement loops

slot out in floor slab for tie steel

reinforcement loops and tie reinforcement

tie steel between slabs

rebated horizontal joint

reinforcement loops and tie steel in vertical joint

precast reinforced concrete floor slab

vertical joint dry packed with concrete

storey height window wall panel

pocket and plate for levelling bolt

Precast concrete wall frame

Fig. 120

sion columns are connected either with studs protruding from column ends and bolted to a connection plate, or by means of a joggle connection.

Composite construction

Composite construction is the name given to structural systems in which the separate structural characteristics and advantages of structural steel sections and reinforced concrete are combined as, for example, in the inverted 'T' beam system described below.

A steel frame, cased in concrete, and designed to allow for the strength of the concrete in addition to that of the steel is a form of composite construction. It has been accepted for some time now that it is reasonable to allow for the stiffening and strengthen-

112

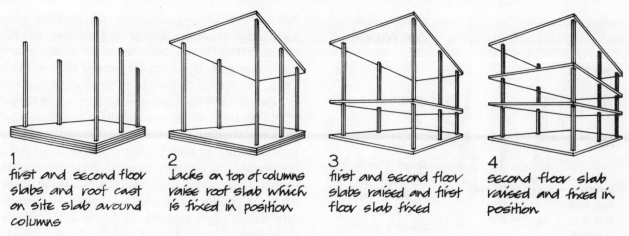

1 first and second floor slabs and roof cast on site slab around columns

2 Jacks on top of columns raise roof slab which is fixed in position

3 first and second floor slabs raised and first floor slab fixed

4 second floor slab raised and fixed in position

Sequence of lifting slabs for a three storey Lift Slab building

Fig. 121

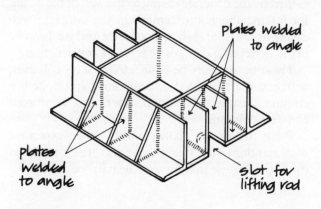

plates welded to angle

plates welded to angle

slot for lifting rod

Lifting collar for lift slab

Fig. 122

ing effect of a solid concrete casing to structural steel beams and columns and it is usual practice for engineers to allow for this effect in their calculations. By reinforcing the concrete casing and allowing for its composite effect with the steel frame, a saving in steel and a reduction in the overall size of members can be effected.

Shear studs and connectors

A concrete floor slab bearing on a steel beam may be considered to act with the beam and serve as the beam's compressive flange, as a form of composite construction. This composite construction effect will only work if there is a sufficiently strong bond between the concrete and the steel, to make them act together in resisting shear stresses developed under load. The adhesion bond between the concrete and the top flange of the beam is not generally sufficient and it is usually necessary to fix shear studs or connectors, to the top flange of the beam, which are then cast in the floor slab. The purpose of these studs and connectors is to provide a positive resistance to shear.

Figure 124 is an illustration of typical shear studs and connectors and composite floor and beam construction.

Inverted 'T' beam composite construction

The composite beam and floor construction des-cribed above employs the standard 'I' section beams. The top flange of the beam is not a necessary part of the construction, as the concrete floor slab can be designed to carry the whole of the compressive stress, so that the steel in the top flange of the beam is wastefully deployed. By using an inverted 'T' section member, steel is placed in the tension area and concrete in the compression area, where their charac-teristics are most useful.

A cage of mild steel binders, cast into the beam casing and linked to the reinforcement in the floor slab, serves to make the slab and beam act as a form of composite construction by the adhesion bond of the concrete to the whole of the 'T' section.

Preflex beams

The use of high tensile steel sections for long span beams has been limited owing to the deflection of the

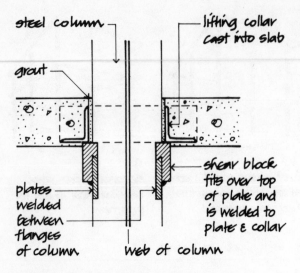

Connection of slab to steel column

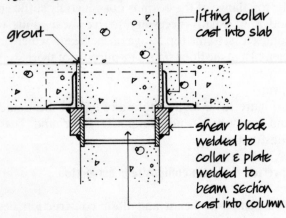

Connection of slab to concrete column

Lift slab – Connection of slab to
columns

Fig. 123

beams under load, which causes cracking of concrete casing, and possible damage to partitions and finishes. Preflex beams are made by applying and maintaining loads, which are greater than working loads, to pairs of beams. In this deflected position, reinforced concrete is cast around the tension flanges of the beams. When the concrete has developed sufficient strength the load is released and the beams tend to return to their former position. In so doing the beams induce a compressive stress in the concrete around the tension flange which prevents the beams wholly regaining their original shape. The beams now have a slight upward camber. Under loads the deflection of these beams will be resisted by the compressive stress in the concrete around the bottom flange which will also prevent cracking of concrete. Further stiffening of the beam to reduce deflection is gained by the concrete casing to the web of the beam. By linking the reinforcement in the concrete web casing to the floor slab, the concrete and steel can be made to act as a composite form of construction.

These beams may be connected to steel columns, with end plates welded to beam ends and bolted to column flanges, or may be cast into reinforced concrete columns.

Prefix beams are considerably more expensive than standard mild steel beams and are designed, in the main, for use in long span heavily loaded floors.

114

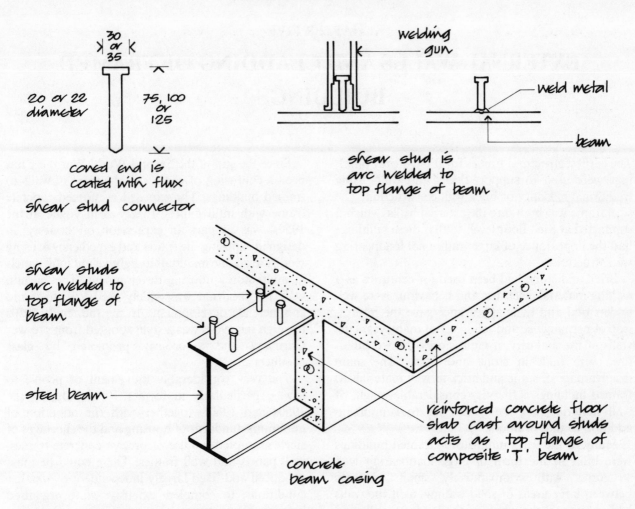

30 or 35

20 or 22 diameter

75, 100 or 125

coned end is coated with flux

shear stud connector

welding gun

weld metal

beam

shear stud is arc welded to top flange of beam

shear studs arc welded to top flange of beam

steel beam

concrete beam casing

reinforced concrete floor slab cast around studs acts as top flange of composite 'T' beam

Shear studs for composite construction

Fig. 124

115

CHAPTER FIVE

EXTERNAL WALLS AND CLADDING OF FRAMED BUILDINGS

The earliest structural frames for multi-storey buildings were used to support floors and walls using traditional masonry or brick wall construction. The steel frame was built into the external walls, which it supported at each floor level, so that these buildings had the appearance of large traditional loadbearing wall structures.

Brick and stone had been used for centuries as a walling material, their use and behaviour were well understood and their appearance gave the impression of permanence and solidity. For many years the walls of the majority of multi-storey framed buildings were built in stone and brick. The main disadvantage of stone and brick as wall materials to framed buildings is the very considerable weight of wall material that the frame has to support in addition to that of floors.

Many of the early multi-storey framed buildings were built in the form of large loadbearing wall structures with comparatively small windows between large areas of solid walling, as if the walls had to support the whole of the loads of walls and floors for the full height of the building. At the time the need for daylight penetration to interiors was subordinate to considerations of appearance. The move towards an increase in the relative area of glass to solid walling to framed structures was gradual and erratic. The acceptance of a radical change of style that has come to be known as the 'modern movement', vigorously promoted in Germany in the 1930s, together with the adoption of the reinforced concrete frame was the cause of a fundamental change in the concept of both the function and appearance of walls.

There was a very gradual and reluctant abandonment of the concept of a wall as a solid enclosure perforated with windows, in favour of the acceptance of the logic of a structural frame supporting a lightweight enclosing weathershield. It was affected by fashion in the form of changes and variations in styles of architecture – the adoption of the concept of what is known as 'building science' and of expediency.

Since the end of the Second World War there has been a confusion of change in the form of walls to framed buildings. The exposed reinforced concrete frame with infill panels, that was in vogue in the 1950s, was in part an expression of 'honesty' in design in exposing the frame and expediency in using readily available materials as lightweight infill panels in multi-storey housing development. The 'curtain wall' of glass that was widely used for office and commercial development from the mid 1950s through the 1960s was a style adopted from pre-war Germany and vigorously promoted by glass producers.

The very considerable movement of people to towns, particularly in former Russia, during the 1950s and 1960s, together with the adoption of minimum standards of housing and the shortages of steel, prompted the use of precast concrete frames, wall panels and wall frames. These could be mass produced and fixed largely independent of weather conditions, to complete extensive state organised housing programmes in most northern European countries. The new wonder material of the petrochemical industry, plastics, reinforced with glass fibre and also, cement, reinforced with glass fibre have to an extent been used as lightweight panels in place of precast concrete.

In the last twenty years, brick has again gained popularity as a material for the walls of framed buildings. The many virtues of this traditional material have been successfully promoted by the Brick Development Association. Many recently completed buildings are clad in brickwork supported by or attached to structural frames, with the brick walling simulating traditional loadbearing forms of piers and arches. Others have employed brick as a facing to such features as projecting canopies in the 'look no hands' abandonment of sense and reality.

More recently, thin sheet metal wall panels have come into vogue, used by themselves as a wall finish or with the frame and services exposed in what is often referred to as 'high tech' architecture. The advantage of a 'rain screen' is being promoted in the use of an outer skin, usually of sheet metal, as an outer protective skin to a backing of weather seal and insulating panels.

The structural frame has provided the possibility of endless variation in the form and appearance of buildings that no longer need be contained inside a loadbearing envelope. Any one of the wall materials that have been used with some degree of success are available to meet the changing needs of use, economy and fashion.

An external wall is the continuous, usually vertical structure that encloses a building to provide privacy, security and equable conditions for comfort, storage and operation for the occupants, goods or machinery housed in the building.

The external walls of framed buildings differ from traditional loadbearing walls because the structural frame has an effect on and influences the design of the wall structure that it supports. To the extent that the structural frame may affect the functional requirements of an external wall, it should be considered as part of the wall structure. The use of the various materials that may be used for the external walls of framed buildings is to an extent influenced by the relative behaviour of the structural frame and the wall to accommodate differential structural, thermal and moisture movements that affect the functional requirements of a wall.

Movements of structural frames

Under load, both steel and concrete structural frames suffer elastic strain and consequent deflection (bending) of beams and floors and shortening of columns. Deflection of beams and floors is generally limited to about one three hundredth of span, to avoid damage to supported facings and finishes. Shortening of columns by elastic strain under load is of the order of 2.5 mm for each storey height of about 4 metres. Elastic shortening of steel columns may be of the order of 1.5 mm per storey height.

The comparatively small deflection of beams and shortening of columns under load can, by and large, be accommodated by the joints in materials such as brick, stone and block and the joints between panels, without adversely affecting the function of most wall structures.

Unlike steel, concrete suffers drying shrinkage and creep in addition to elastic strain. Drying shrinkage occurs as water, necessary for the placing of concrete and setting of cement, migrates to the surface of concrete members. The rate of loss of water and consequent shrinkage depend on the moisture content of the mix, the size of the concrete members and atmospheric conditions. Drying shrinkage of concrete will continue for some weeks after placing. For the small members of a structural frame, drying out of doors in summer, about half of the total shrinkage takes place in about one month and about three quarters in six months. For larger masses of concrete, about half the total shrinkage takes place one year after placing.

Drying shrinkage of concrete is restrained by the bond between concrete and reinforcement to the extent that concrete in heavily reinforced members shrinks less than that in lightly reinforced sections. Drying shrinkage of the order of 1.2 mm for each 4 metres of column length may well occur.

Creep of concrete is dependent on stress and is affected by humidity and by the cement content and the nature of the aggregate in concrete. The gradual creep of concrete may continue for some time and result in a shortening of columns of the order of up to 2.5 mm for each storey height of column over the long term. Like drying, shrinkage creep is restrained by reinforcement.

The combined effect of elastic strain, drying shrinkage and creep in concrete may well amount to a total reduction of up to 6 mm for each storey height of building. Because of these effects, it is necessary to make greater allowance for shortening in the design of wall structures supported by an in situ cast concrete frame than it is for a steel frame. Solid wall structures, such as brick, which are built within or supported by a concrete structural frame should be built with a 12 to 15 mm compression joint at each floor level to avoid damage to the wall by shortening of the frame and expansion of the wall materials due to thermal and moisture movements.

Experience shows that there are generally considerably greater inaccuracies in line and level with in situ cast concrete frames than there are with steel frames. There is an engineering tradition of accuracy of cutting and assembling steel that is not matched by the usual assembly of formwork for in situ cast concrete. Deflection of formwork under the load of wet concrete and some movement of formwork during the placing and compaction of concrete combine to create inaccuracies of line and level of both beams and columns in concrete frames that may be magnified up the height of multi-storey buildings. Allowances for these inaccuracies can be made where fixings for cladding are made by drilling for bolt fixings rather than relying on cast on or cast in supports and fixings. The advantage of the precast concrete frame is in the greater accuracy of casting of concrete in controlled factory conditions than on site.

FUNCTIONAL REQUIREMENTS

The functional requirements of a wall are:

Strength and stability
Resistance to weather
Durability and freedom from maintenance
Fire safety
Resistance to the passage of heat
Resistance to the passage of sound

Strength and stability

A wall structure should have adequate strength to support its own weight between points of support or fixing to the structural frame, and sufficient stability against lateral wind pressures.

To allow for differential movements between the structural frame and the wall structure there has to be adequate support to carry the weight of the wall structure, and also restraint fixings that will maintain the wall in position and at the same time allow differential movements without damage to either the fixings or the wall material.

Brick and precast concrete cladding do not suffer the rapid changes of temperature between day and night that thin wall materials do, because they act to store heat and lose and gain heat slowly. Thin sheet wall materials such as GRP, metal and glass suffer rapid changes in temperature and consequent expansion and contraction which may cause distortion and damage to fixings or the thin panel material or both.

In the design of wall structures faced with thin panel or sheet material the ideal arrangement is to provide only one rigid support fixing to each panel or sheet with one other flexible support fixing and two flexible restraint fixings. The need to provide support and restraint fixings with adequate flexibility to allow for thermal movement and at the same time adequately restrain the facing in place and maintain a weathertight joint, has been the principal difficulty in the use of thin panel and sheet facings.

Resistance to weather

The traditional walling materials, brick, stone and block, serve to exclude rain from the inside of buildings by absorbing rain water that evaporates to outside air during dry periods. The least thickness of solid wall material necessary to prevent penetration of rain water to the inner face depends on the degree of exposure to driving rain (see Volume 1). Common practice is to construct walls as a cavity wall with an outer leaf of brick as rain screen, a cavity and an inner leaf of some lightweight block as thermal barrier and solid inside wall surface.

Individual precast concrete wall panels act in much the same way as brick by absorbing rain water. Because of the considerable size of these panels there have to be comparatively wide joints between panels to accommodate structural, thermal and moisture movements. The joints are designed with a generous overlap to horizontal and an open drained joint to vertical joints to exclude rain.

Non-absorbent sheet materials, such as metal and glass, cause driven rain to flow under pressure in sheets across the face of the wall so making the necessary joints between panels of the material highly vulnerable to penetration by rain. These joints should at once be sufficiently wide to accommodate structural, thermal and moisture movements and serve as an effective seal against rain penetration. The materials that are used to seal joints are mostly short-lived as they harden on exposure to atmosphere and sun and lose resilience in accommodating movement.

The 'rain screen' principle is designed to provide a separate outer skin, to screen wall panels from scouring by wind and rain and deterioration by sunlight and to improve the life and efficiency of joint seals.

Durability and freedom from maintenance

The durability of a wall structure is a measure of the frequency and extent of the work necessary to maintain minimum functional requirements and acceptable appearance.

Walls of brick and natural stone will very gradually change colour over the years. This slow change of colour, termed weathering, is generally accepted as one of the attractive features of these traditional wall materials. Walls of brick and stone facing require very little maintenance over the expected life of most buildings. Precast concrete wall panels which weather gradually may become dirt stained due to slow run off of water from open horizontal joints. This irregular and often unsightly staining, particularly around top edges of panels, is a consequence of the panel form of this type of cladding.

Panels of glass will maintain their lustrous fire glazed finish over the expected life of buildings, but will require frequent cleaning of the surface, if they are to maintain their initial appearance, and periodic attention to and renewal of seals.

Of the sheet metal facings that can be used for wall structures, bronze and stainless steel, both expensive materials, will weather by the formation of a thin film of oxide that is impermeable and prevents further oxidation. Aluminium which weathers with a light coloured, coarse textured oxide film that considerably alters the appearance of the surface can be anodised to inhibit the formation of an oxide film or coated with a plastic film for the sake of appearance. Steel, which progressively corrodes to form a porous oxide, is coated with zinc to inhibit the rust formation and a plastic film as decoration.

None of the plastic film coatings are durable (see Volume 3) as they lose colour over the course of a few years on exposure to sunlight and this irregular colour bleaching may well not be acceptable from the point of view of appearance to the extent that painting or replacement may be necessary in 10 to 25 years. In common with other thin panel materials there will be a need for periodic maintenance and renewal of seals to joints between metal faced panels.

Fire safety

The requirements from Part B of Schedule 1 to the Building Regulations 1991, as amended 1994, are concerned to:

(a) provide adequate means of escape
(b) limit internal fire spread (linings)
(c) limit internal fire spread (structure)
(d) limit external fire spread
(e) provide access and facilities for the Fire Services

Fire safety regulations are concerned to ensure a reasonable standard of safety in case of fire. The application of the regulations, as set out in the practical guidance given in Approved Document B, is directed to the safe escape of people from buildings in case of fire rather than the protection of the building and its contents. Insurance companies that provide cover against the risks of damage to the buildings and contents by fire will generally require additional fire protection such as sprinklers.

Means of escape

The requirement from the Building Regulations is that the building shall be designed and constructed so that there are means of escape from the building in case of fire to a place of safety outside the building. The main dangers to people in buildings, in the early stages of a fire, are the smoke and noxious gases produced which cause most of the casualties and may also obscure the way to escape routes and exits. The Regulations are concerned to:

(a) provide a sufficient number and capacity of escape routes to a place of safety
(b) protect escape routes from the effects of fire by enclosure, where necessary, and to limit the ingress of smoke
(c) ensure the escape routes are adequately lit and exits suitably indicated

The general principle of means of escape is that any person in a building confronted by an outbreak of fire can turn away from it and make a safe escape.

The number of escape routes and exits depends on the number of occupants in the room or storey, and the limits on travel distance to the nearest exit depend on the type of occupancy. The number of occupants in a room or storey is determined by the maximum number of people it is designed to hold, or calculated by using a floor space factor related to the type of accommodation, which is used to determine occupancy related to floor area as set out in Approved Document B. The maximum number of occupants determines the number of escape routes and exits; where there are no more than 50 people 1 escape route is acceptable. Above that number, a minimum of 2 escape routes is necessary for up to 500 and up to 8 for 16 000 occupants. Maximum travel distances to the nearest exit are related to purpose-group types of occupation and whether one or more escape routes are available. Distances for one direction escape are from 9.0 to 18.0 and for more than one direction escape from 18.0 to 45.0, depending on the purpose groups defined in Approved Document B.

Internal fire spread (linings)

Fire may spread within a building over the surface of materials covering walls and ceilings. The Regulations prohibit the use of materials that encourage spread of flame across their surface when subject to intense radiant heat and those which give off appreciable heat when burning. Limits are set on the use of thermoplastic materials used in rooflights and lighting diffusers.

Internal fire spread (structure)

As a measure of ability to withstand the effects of fire, the elements of a structure are given notional fire resistance times, in minutes, based on tests. Elements are tested for the ability to withstand the effects of fire in relation to:

(a) resistance to collapse (loadbearing capacity) which applies to loadbearing elements
(b) resistance to fire penetration (integrity) which applies to fire separating elements
(c) resistance to the transfer of excessive heat (insulation) which applies to fire separating elements

The notional fire resistance times, which depend on the size, height and use of the building, are chosen as being sufficient for the escape of occupants in the event of fire.

The requirements for the fire resistance of elements of a structure do not apply to:

(1) A structure that only supports a roof unless
  (a) the roof acts as a floor, e.g. car parking, or as a means of escape
  (b) the structure is essential for the stability of an external wall which needs to have fire resistance.
(2) The lowest floor of the building.

Compartments

To prevent rapid fire spread which could trap occupants, and to reduce the chances of fires growing large, it is necessary to subdivide large buildings into compartments separated by walls and/or floors of fire-resisting construction. The degree of subdivision into compartments depends on:

(a) the use and fire load (contents) of the building
(b) the height of the floor of the top storey as a measure of ease of escape and the ability of fire services to be effective
(c) the availability of a sprinkler system which can slow the rate of growth of fire

The necessary compartment walls and/or floors should be of solid construction sufficient to resist the penetration of fire for the stated notional period of time in minutes. The requirements for compartment walls and floors do not apply to single storey buildings.

Concealed spaces

Smoke and flames may spread through concealed spaces, such as voids above suspended ceilings, roof spaces and enclosed ducts and wall cavities in the construction of a building. To restrict the unseen spread of smoke and flames through such spaces, cavity barriers and stops should be fixed as a tight fitting barrier to the spread of smoke and flames.

External fire spread

To limit the spread of fire between buildings, limits to the size of 'unprotected areas' of walls and also finishes to roofs, close to boundaries, are imposed by the Building Regulations. The term 'unprotected area' is used to include those parts of external walls that may contribute to the spread of fire between buildings. Windows are unprotected areas as glass offers negligible resistance to the spread of fire. The Regulations also limit the use of materials of roof coverings near a boundary that will not provide adequate protection against the spread of fire over their surfaces.

Access and facilities for the Fire Services

Buildings should be designed and constructed so that:

- Internal firefighting facilities are easily accessible
- Access to the building is simple
- Vehicular access is straightforward
- The provision of fire mains is adequate

Resistance to the passage of heat

The interior of buildings built with insulated solid masonry walling and those clad with insulated panels

120

of concrete, GRC, GRP, glass and metal, is heated by the transfer of heat from heaters and radiators to air (conduction), the circulation of heated air (convection) and the radiation of energy from heaters and radiators to surrounding colder surfaces (radiation). This internal heat is transferred to and through colder enclosing walls, roof and floors by conduction, convection and radiation to colder outside air.

The interior of buildings clad with large areas of glass may gain a large part or the whole of their internal heat from a combination of solar heat gain through glass cladding and from internal artificial lighting to the extent that there may be little need for supplementary internal heating. As long as the interior of buildings is heated to a temperature above that of outside air, transfer of heat from heat sources to outside air will continue. For the sake of economy in the use of expensive fuel and power sources, and to conserve limited supplies of fuel, it is sensible to seek to limit the rate of transfer of heat from inside to outside. Because of the variable complex of the modes of transfer of heat it is convenient to distinguish three separate modes of heat transfer as conduction, convection and radiation.

Conduction is the direct transmission of heat by contact between particles of matter, convection the transmission of heat by the motion (circulation) of heated gases and fluids, and radiation the transfer of heat from one body of radiant energy through space to another by a motion of vibration in space which radiates equally in all directions.

Conduction

The speed or rate at which heat is conducted through a material depends mainly on the density of the material. Dense metals conduct heat more rapidly than less dense gases. Metals have high and gases have low conductivity. Thermal conductivity (λ-value) is the rate of heat per unit area conducted per unit time through a slab of unit thickness per degree of temperature difference. It is expressed in watts per metre thickness of material per kelvin (W/mK) where W (watt) is the unit of power which is equivalent to joules (the unit of heat) per second (J/s) and the temperature is expressed in kelvin (K).

Convection

The density of air that is heated falls and the heated air rises and is replaced by cooler air. This, in turn, is heated and rises so that there is a continuing movement of air as heated air loses heat to surrounding cooler air and cooler surfaces of ceilings, walls and floors. Because the rate of transfer of heat from air to cooler surfaces varies from rapid transfer through thin sheets of glass in windows to an appreciably slower rate of transfer to insulated walls by conduction, and because of the variability of the exchange of cold outside air with warm inside air by ventilation, it is not possible to quantify heat transfer by convection. Usual practice is to make an assumption of likely total air changes per hour or volume (litres) per second depending on categories of activity in rooms and then to calculate the heat required to raise the temperature of the fresh, cooler air introduced by natural or mechanical ventilation, making an assumption of the temperature of inside and outside air.

Radiation

Energy from a heated body radiating equally in all directions is partly reflected and partly absorbed by another cooler body (with the absorbed energy converted to heat). The rate of emission and absorption of radiant energy depends on the temperature and the nature of the surface of the radiating and receiving bodies. The heat transfer by low temperature radiation from heaters and radiators is small whereas the very considerable radiant energy from the sun that may penetrate glass and that from high levels of artificial illumination is converted to appreciable heat inside buildings. An estimate of the solar heat gain and heat gain from artificial illumination may be assumed as part of the heat input to buildings and used in the calculation of heat input and loss.

Transmission of heat

The practical guidance in Approved Document L to meeting the requirements from Part L of Schedule 1 to the Building Regulations 1991, as amended 1994, for the conservation of fuel and power, is mainly directed to limiting the loss of heat through the fabric (walls, floors and roofs) of buildings, other than dwellings, by establishing maximum values for the overall transmission of heat, the 'U' value, through walls, roofs and floors and to limiting the size of glazed areas.

Because of the complexity of the combined modes of transfer of heat through the fabric of buildings it is convenient to use a coefficient of heat transmission, the U value. This air-to-air thermal transmittance coefficient, the U value, takes account of the transfer of heat by conduction through solid materials and gases, convection of air in cavities and across inside and outside surfaces and radiation to and from surfaces. The U value expressed as W/m^2K is a measure of how much heat will pass through one square metre of a structure when the combined radiant and air temperature on each side of the structure differ by $1K$. A high U value indicates comparatively high rates of overall transmission and a low U value indicates low rates.

The three methods of showing compliance with the regulation for conservation of fuel and power for all other buildings than dwellings, which are set out in Approved Document L, are:

- an elemental method
- a calculation method
- an energy use method

The *elemental method* relates thermal performance to a table of standard U values for the envelope of buildings and a basic allowance for windows, doors and rooflights as a percentage of exposed wall area, and as a percentage of roof area for rooflights.

The standard U values given in Approved Document L are $0.45 W/m^2K$ for exposed walls, floors, ground floors and $0.25 W/m^2K$ for roofs of buildings other than dwellings.

The loss of heat through windows, doors and rooflights is limited by setting maximum sizes related to floor and roof areas as:

Windows
40% of exposed wall areas for places of assembly, offices and shops
15% of exposed wall area for industrial and storage.
Rooflights
20% of roof area for all buildings.

A modification of the basic allowance for windows, doors and rooflights, which is based on a U value of $3.3 W/m^2K$ may be made where the U value is other than that used in the basic allowance.

Appendix A of Approved Document L provides a series of tables setting out the base thickness of insulation required in typical roof, floor and wall construction necessary to meet the requirements for thermal performance.

The *calculation method* allows, within certain limits, greater flexibility in selecting areas of windows, doors and rooflights and/or the level of insulation of individual elements of the building envelope than the elemental method does, providing the calculated rate of heat loss through the envelope is not greater than that of a building of the same size and shape designed to comply with the elemental method.

The *energy use method* allows completely free design using any valid conservation measure and takes account of useful solar and internal heat gains, providing the calculated annual energy use of the proposed building is less than the calculated annual energy use of a similar building designed to comply with the elemental method.

In the calculation and energy use methods maximum U values are set at $0.45 W/m^2K$ for the roof and $0.7 W/m^2K$ for the walls and floors of residential buildings and $0.7 W/m^2K$ for the roof, walls and floors of non-residential buildings.

In the calculation and energy use methods for buildings other than dwellings there is no requirement to use the standard assessment procedure for calculating the rate of heat loss or calculation of annual energy use.

Buildings which are clad with large areas of glass as a wall envelope can meet the requirements for the conservation of energy by taking account of solar heat gain through surface modified or surface coated solar control glass. The effect of this type of glass is that the surface coating reflects back into the building the long wave energy generated by heating, lighting and occupants, while permitting the transmission of short wave energy from outside (see Volume 2).

Condensation

During recent years increased expectation of thermal comfort in buildings and the need to conserve limited supplies of fuel and power has led to improved levels of insulation in the fabric of buildings and the common use of weatherseals to opening windows that has restricted natural ventilation. These changes have led to the likelihood of increased levels of humid conditions that cause condensation on the inner faces of cold surfaces such as glass in windows and the inside of thin metal sheet weathering.

The limited capacity of air to take up water in the form of water vapour increases with temperature so that the warmer the air, the greater the amount of water vapour it can hold. The amount of water vapour held in air is expressed as a ratio of the actual amount of water vapour in the air to the maximum which the air could contain at a given temperature. This relative humidity (rh) is given as a percentage. Air is saturated at 100% relative humidity and the temperature at which this occurs is defined as the dew point temperature. When the temperature of warm moist air falls to a temperature at which its moisture vapour content exceeds the saturation point, the excess moisture vapour will be deposited as water. This will occur, for example, where warm moist air comes into contact with cold window glass, its temperature at the point of contact with the glass falls below that of its saturation point and the excess moisture vapour forms as droplets of water on the inside window surface as condensation. Thus, the greater the amount of moisture vapour held in the air and the greater the temperature difference between the warm inside air and the cold window surface, the more the condensation.

The main sources of moisture vapour in air in buildings are moisture given off by occupants, moisture given off from cooking and flueless heaters, bathing, clothes washing and drying, moisture generating processes and the drying out from a new building.

Internal temperatures which are high relative to cold outside air will tend to produce high levels of condensation on cold surfaces from high levels of moisture vapour. An atmosphere that contains high levels of moisture vapour is said to be humid. The level of humidity that is acceptable for comfort in buildings varies from about 30% to 70% relative humidity. Low levels of humidity, below about 20%, may cause complaints of dry throats and cause woodwork to shrink and crack. High levels of humidity, e.g. above 70%, may cause discomfort and lead to condensation on cold surfaces, mould growth and excess heat. For comfort in buildings and to limit the build-up of humidity it is necessary to provide a degree of ventilation for an adequate supply of oxygen and to limit fumes, body odour and smells and to exchange drier fresh air from outside with humid, stale inside air. The level of ventilation required depends on the activity and number of people in a given space and sources of heat and water that will contribute to increased humidity.

Ventilation

The use of sealed glazing and effective weatherseals to the joints of cladding panels and windows in the envelope of modern buildings has restricted the natural exchange of outside and inside air to provide ventilation of buildings. For comfort there should be a continuous change of air inside buildings to provide an adequate supply of oxygen, to limit the build-up of humidity, fumes, body odour and smells and provide a regular movement of air that is necessary for bodily comfort. The necessary movement of air inside sealed buildings may be induced artificially by mechanical systems of air conditioning which filter, dry and humidify air through a complex of inlet and extract ducting, connected to one or more air treatment plants. The pumps necessary to force air through the ducts may cause an unacceptable level of noise and the air handling system is costly to instal, maintain and run. To economise, it is often practice to install individual air conditioning heaters which filter, dry and heat air that is recirculated from individual rooms with the effect that stale air is constantly circulated so causing conditions of discomfort.

As an alternative, buildings may be constructed and finished with mainly open plan floor areas largely free of enclosed spaces, set around one or more central areas open from ground to roof level to provide facility for the natural movement of heated air to rise and so cause natural ventilation. This stack system of ventilation, so called by reference to the upward movement of air up a chimney stack, can be utilised by itself or with some small mechanical ventilation to provide comfort conditions with the least initial and running costs.

Thermal bridge (cold bridge)

Thermal bridge is the term used to describe a material of high conductivity, that is a poor thermal insulator, in a wall structure that offers little resistance, or appreciably less resistance, to the transfer of heat than the surrounding material or materials. The prime example of a thermal bridge is the thin glass in windows which offers negligible resistance to transfer of heat and serves as a bridge to the transfer of cold from outside (cold bridge) or heat from outside (hot bridge) to the inside of buildings.

The members of a structural frame act as a thermal bridge where the wall is built up to or between the frame as is common with solid wall structures, as illustrated later in Fig. 126 where the resistance to the thermal transfer through the brick slips and beam is appreciably less than through the rest of the wall. Similarly there is a thermal bridge across a precast wall panel and a beam and column as illustrated later in Fig. 149.

A thermal bridge will also encourage condensation of warm moist air on its cold surfaces as, for example, where there is a cold bridge in the external wall of a bathroom or kitchen, on which water vapour will condense from the warm moisture laden inside air and cause staining and possible damage to finishes.

There is no single or several wholly effective ways of preventing the thermal bridge formed by the supporting structural frame. The effect of the bridge may be modified by the use of floor insulation and a suspended ceiling or by setting frame members, where possible, back from the outer face of the wall, as illustrated later in Fig. 127.

Wall panels of precast concrete, GRP and GRC have been used with a sandwich or inner lining of an insulating material. This arrangement is not entirely effective because the insulating material, if open pored as are many insulating materials, may absorb condensate water which will reduce its thermal properties and the edge finish to panels, necessary for rigidity and jointing, will act as a thermal bridge, as illustrated later in Fig. 155.

Thin metal wall panel materials which are supported by a metal carrier system fixed across the face of the structural frame can provide thermal insulation more effectively by a sandwich, inner lining or inner skin of insulating material with the edge jointing material acting as a thermal break in the narrow thermal bridge of the edge metal, as illustrated later in Fig. 180.

Resistance to the passage of sound
(see also Volume 2)

Sound is the effect of air disturbances that radiate from a source as 'airborne sound' or by a vibration in solid material as 'impact sound'. The awareness of sound depends on the background level to the extent that where the background level is high, as in cities, only comparatively loud sounds are intrusive, whereas in rural areas where background sound levels are low, a small sound may be intrusive.

Sound from a source such as a radio is generated by a cyclical disturbance of air that radiates from the source. The most effective barrier to airborne sound is an intervening mass such as a solid wall. The more dense and thick the material, the more effective it is as a barrier to airborne sound as the dense mass absorbs the energy generated by the sound source.

Impact sound is generated when a solid material transfers the energy by vibration. The continuous solid material of a structural frame is a prime conductor of impact sound. The sound of a door slammed shut may be transferred audibly through several floors of a structural frame where background levels of sound are low. Unexpected impact sounds can be most disturbing.

The most effective way of reducing impact sound is to isolate the potential source of impact from continuous solid transmitters such as structural frames. Resilient fixings to door frames and resilient bushes to supports for hard floor finishes effectively isolate the source of common impact sounds.

EXTERNAL WALLS AND CLADDING

The words wall, walling, cladding, facings and wall facings are variously used relative to the usually vertical envelope of buildings.

By definition a wall is a continuous, usually vertical structure of brick, stone, block, concrete, timber or metal, thin in proportion to its height and length. The traditional wall of stone or brick, which is built off a foundation, is self-supporting to the extent that the whole of the weight of the wall, and such floors and roofs as it supports, is carried by the foundation.

The word cladding came into general use as a description of the external envelope of framed buildings, which clothed or clad the building in a protective coating that was hung, supported by or secured to the skeleton or structural frame as a jacket or coat is hung from the shoulders as a protective cladding.

The word facings has been used to describe materials used as a thin, non-structural, decorative, external finish such as the thin, natural stone facings applied to brick or concrete backing.

The words wall or walling will be used to describe the use of those materials such as stone, brick, concrete and blocks that are used as the external envelope of framed buildings where the appearance is of a continuous wall to the whole or part of several storeys or as walling between exposed, supporting beams and columns of the frame.

The word cladding will be used to describe panels of concrete, GRC, GRP, glass and metal fixed to and generally hung from the frame by supporting beams or inside light framing as a continuous outer skin to the frame.

The external walls and cladding of framed buildings are broadly grouped as:

- Solid and cavity walling of stone, block and brick
- Facings applied to solid and cavity background walls
- Cladding panels of precast concrete, GRC and GRP
- Infill wall framing to a structural grid
- Thin sheet cladding of glass and metal

SOLID AND CAVITY WALLING

Natural and reconstructed stone walling

The traditional use of ashlar stone walling (see Volume 1), of accurately squared stones for the external face of large buildings, continued in the construction of many of the earliest framed buildings, such as the Ritz Hotel in London, for the sense of solidity and permanence that such a finish provided. The solid walls faced with ashlar natural stone with a backing of squared stone or brick were built around the supporting steel columns and supported at each floor level by steel beams built in to support some two-thirds of the wall thickness at each floor level. Large projecting cornice stones were supported by cantilevered steel beams around which the stones were cut to fit at vertical joints.

The advantage of the supporting steel frame was a reduction in the thickness of the walling, particularly in the lower storeys of multi-storey buildings, as compared to a fully self-supporting wall of similar height and the improved resistance to damage by fire of the steel frame by virtue of the enclosing walls. Many of the earlier multi-storey buildings in the USA were constructed this way for the increase in floor area as compared to traditional construction and thus increased rental value of the building, where taxes were levied on site area.

In London and other cities in England, terra cotta (burnt earth) blocks were much used in the walls of large buildings built in the Victorian period. Fired blocks of terra cotta, with a semi-glaze finish, were moulded in the form of natural stone blocks, both plain and heavily ornamented, to replicate the form and detail of the stonework buildings of the time. The plain and ornamented blocks were made hollow to reduce and control shrinkage of wet clay during firing. In use the hollows in these blocks were filled with concrete and the blocks then laid as if they were natural stone. Well-burnt blocks of terra cotta are durable even in heavily polluted atmospheres where natural limestone and sandstone facings deteriorate.

The massive form of construction of stone or terra cotta walling to framed buildings, which imposed very considerable loads on the supporting frame and foundations and was needlessly extravagant in the use of costly materials, provided adequate resistance to weather, poor resistance to the passage of heat and allowance for such structural, thermal and moisture movements as might occur, in the many joints in the walling material and the massive construction. To economise in the use of limited supplies of expensive stone it was practice to bond ashlar facing stones to a backing of roughly squared stone or brickwork.

More recently to improve thermal resistance and to provide a cavity as a barrier to the penetration of water to the inner face, it became practice to construct stone faced walling as a cavity wall with the outer leaf of stone bonded to a backing of brick, a cavity and an inner leaf of lightweight block similar to the loadbearing wall illustrated in Volume 1. The frame provided support for the outer leaf through beams built in at each floor level to support two-thirds of the thickness of the outer leaf. This form of construction provided appreciable resistance to the transfer of heat and improvement to the resistance to the penetration of water to the inner face and some little reduction in loads on the frame and foundations. For reasons of economy in labour and the use of the limited supply of native natural stone solid or cavity stone faced buildings are rarely constructed with solid stone walling today.

Reconstructed stone

Reconstructed stone is one of the several terms that have been used to describe a wall material that is made with aggregate and cement to resemble in appearance and be used in a similar way to natural stone. When the material first came into use some fifty years ago it was called artificial stone. To avoid the use of the word artificial, the material is generally referred to as reconstructed stone. The British Standard refers to the material as cast stone, which is a fair description of the method of manufacture. The majority of manufacturers of the material prefer the description reconstructed stone.

The material is made by casting a mix of natural stone aggregate, cement and water inside timber moulds by either the moist earth or the plastic or wet mix method. The moist earth method of mixing and casting is used principally for both plain and ornamental stones similar in size to natural stone ashlar blocks, where the fine natural stone aggregate and the comparatively dry mix produces a finish that very closely resembles that of the natural stone from which the aggregate is crushed, to the extent that it is often difficult to distinguish reconstructed from natural stone. Natural stone, such as Portland or Bath, crushed to a maximum size of 6 or less, is mixed with cement and just sufficient water that the material has the consistency and feel of moist earth in the hand. This mix is spread in the bed of the mould to a thickness of about 30 and compacted to a thickness of about 20 by electric or air-powered hammering. Ordinary backing concrete is then placed and compacted over the drier mix, in layers of about 50, to the required depth. Where necessary, reinforcement is cast into the backing concrete with the reinforcement at least 30 from the finished face.

The advantage of the moist earth method of casting is that the finished surface closely resembles that of the natural stone from which the face aggregate is produced and the finished surface requires little if any surface finishing. Because of its fine texture the facing material can be carved. More usually today ornamental carved finishes to reconstructed stone blocks are produced by casting on to GRP or rubber formers.

The majority of reconstructed stone blocks used for ashlar and ornamental stone facings that resemble natural limestone and sandstone finishes, are cast by the moist earth method. The main limitation of this method of casting is that return faces are generally limited to about 200.

The plastic or wet mix method of casting reconstructed stone is used for the larger blocks and for cladding panels where deep return faces and ornamentation cannot successfully be cast by the moist earth method. With the plastic method of casting a comparatively wet mix of crushed natural stone aggregate, graded up to 14, is mixed with ordinary or

white cement and poured into the mould inside which it will, because of the wet mix, find its own level during consolidation. The mix, which is consolidated by vibrating tables, will tend to be more dense on the lower faces of the mould. Surface voids on upper faces are filled with the same fine material once the block is taken from the mould. The wet mix is consolidated around the necessary reinforcement, suspended inside the mould. Because of the wet mix that is used it is essential that the moulds be watertight to avoid leakages of grout that might otherwise stain exposed faces.

Because of the consolidation of the wet mix in the mould there is a surface of fine particles of aggregate and cement on finished faces. This fine surface, of what is called laitance, is removed to expose the underlying aggregate cement mix by sand blasting or by wet grinding. Well-made reconstructed stone is generally indistinguishable from the natural stone it is made from and will often be as durable and weather just like the natural material. In the early years when this material was extensively used a number of inferior products were produced which rapidly lost colour, became dirt stained and gave the material, then known as artificial stone, a bad name which it has only recently lost.

Solid and cavity brick walling

The durable, economic, familiar, traditional wall material, brick, has for many years been one of the principal materials used as walling to framed buildings.

In the early days of the multi-storey structural frame, brick walling was built as a large solid loadbearing brick structure with traditional windows and the frame members built in to the walling on each floor level as illustrated in Fig. 125. Beams built into the walling at each floor level provided support

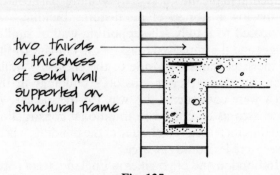

two thirds of thickness of solid wall supported on structural frame

Fig. 125

for two-thirds of the thickness of the wall. The horizontal beams provided adequate support and the massive construction of the wall adequate resistance to lateral wind pressure and suction. At each supporting beam, bats of half a brick were used to mimic the facing headers in the rest of the wall. The considerable thickness of the wall was generally sufficient to resist penetration of rainwater to the inside face. As there was at the time no requirement for the conservation of fuel, the comparatively poor resistance of the wall to the passage of heat was not a consideration. The disadvantages of this form of

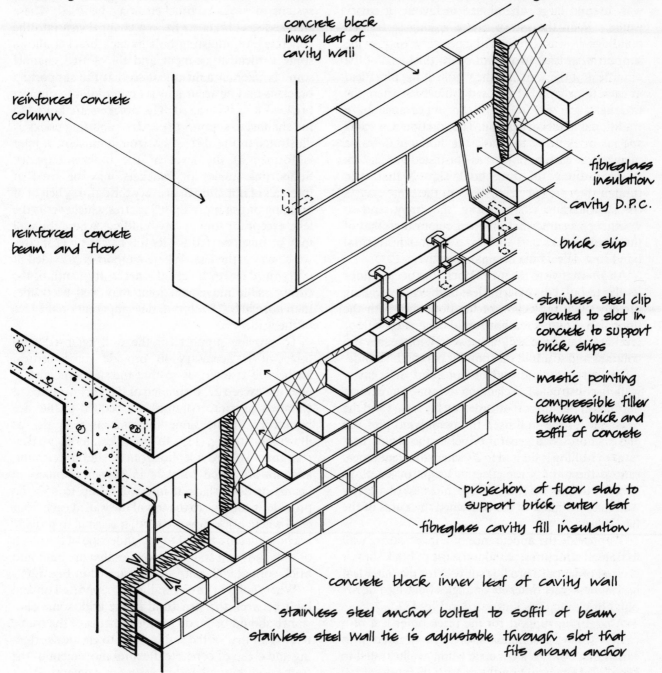

concrete block inner leaf of cavity wall

reinforced concrete column

reinforced concrete beam and floor

fibreglass insulation

cavity D.P.C.

brick slip

stainless steel clip grouted to slot in concrete to support brick slips

mastic pointing

compressible filler between brick and soffit of concrete

projection of floor slab to support brick outer leaf

fibreglass cavity fill insulation

concrete block inner leaf of cavity wall

stainless steel anchor bolted to soffit of beam

stainless steel wall tie is adjustable through slot that fits around anchor

Brick cladding to reinforced concrete frame

Fig. 126

construction were the gross mass of the walling and the cost of the necessary framing to support it and the poor thermal resistance of the wall.

With the introduction of the cavity wall, changing fashion and increasing expectation of thermal comfort in buildings, solid walling to framed buildings was, by and large, abandoned in favour of cavity walling. With the use of cavity walling to framed buildings it was considered necessary to provide support of at least two-thirds of the thickness of the outer leaf of the wall and the whole of the inner leaf at each floor level. This posed difficulties where the external face was to have the appearance of a traditional loadbearing wall. The solution was to fix special brick slips to mask the horizontal frame members at each floor level as illustrated in Fig. 126. A disadvantage of these brick slips is that even though they are cut or made from the same clay as the surrounding whole bricks, they may tend to weather to a somewhat different colour than that of the whole bricks and so form a distinct horizontal band that defeats the original objective.

An alternative to the use of brick slips at each floor level is to build the external leaf of the cavity walling directly off a projection of the floor slab with the floor slab exposed as a horizontal band at each floor level or to build the walling between floor beams and columns and so admit the frame as part of the facade.

The strength and stability of solid and cavity walling constructed as cladding to framed structures depend on the support afforded by the frame and the resistance of the wall itself to lateral wind pressure and suction. As a general principle the slenderness ratio of walling is limited to 27 where the slenderness ratio is the ratio of the effective height or length to effective thickness. The effective thickness of a cavity wall may be taken as the combined thickness of the two leaves.

To provide the appearance of a loadbearing wall to framed structures, without the use of brick slips, it is usual practice to provide support for the outer leaf by stainless steel brackets or angles built into horizontal brick joints as illustrated in Fig. 127.

A common support for the brick outer leaf of a cavity wall is a stainless steel angle secured with expanding bolts to a concrete beam as illustrated in Fig. 127. Depending on the relative thickness of the supporting flange of the angle and the thickness of the mortar joints, the angle may be bedded in the mortar joint or the bricks bearing on the angle may be cut to fit over the angle. To allow for relative movement between walling and the frame it is usual practice to form a horizontal movement joint at the level of the support angle by building in a compressible strip which is pointed on the face with mastic to exclude water.

As an alternative to a continuous angle support a system of support brackets may be used. These stainless steel brackets fit to a channel cast into the concrete. An adjusting bolt in each bracket allows some vertical adjustment and the slotted channel some horizontal adjustment so that the supporting brackets may be accurately set in position to support brickwork as it is raised. The brackets are bolted to the channel to support the ends of abutting bricks as illustrated in Fig. 128. A horizontal movement joint is formed at the level of the bracket support. Supporting angles or brackets may be used at intervals of not more than every third storey height of building or not more than 9 metres whichever is the less, except for four storey buildings where the wall may be unsupported for its full height or 12 metres whichever is the less. Where support is provided at every third storey height the necessary depth of the compressible movement joint may well be deeper than normal brick joints and be apparent on the face of the wall.

To provide support for the wall against lateral forces it is necessary to provide some vertical anchorage at intervals so that the slenderness ratio does not exceed 27. Fishtailed or flat anchors fitted to channels cast into columns are bedded in the face brickwork at the same intervals as wall ties as illustrated in Fig. 129, to provide lateral, vertical restraint. To provide horizontal lateral restraint, anchors are fitted to slots in cast-in channels in beams or floor slabs at intervals of up to 450. To provide anchorage to the top of the wall at each floor level where brick slips are used, it is usual to provide anchors that are bolted to the underside of the beam or slab and to fit stainless steel ties that are built into brickwork at 900 centres as illustrated in Fig. 126.

Where solid or cavity walling is supported on and built between the structural frame grid, some allowance should be made for movements of the frame relative to that of the walling due to elastic shortening and creep of concrete, flexural movement of the frame and thermal and moisture movements. Practice is to build in some form of compressible filler at the junction of the top of the walling and the frame members and the wall and columns as movement joints, with metal anchors set into cast-in channels in

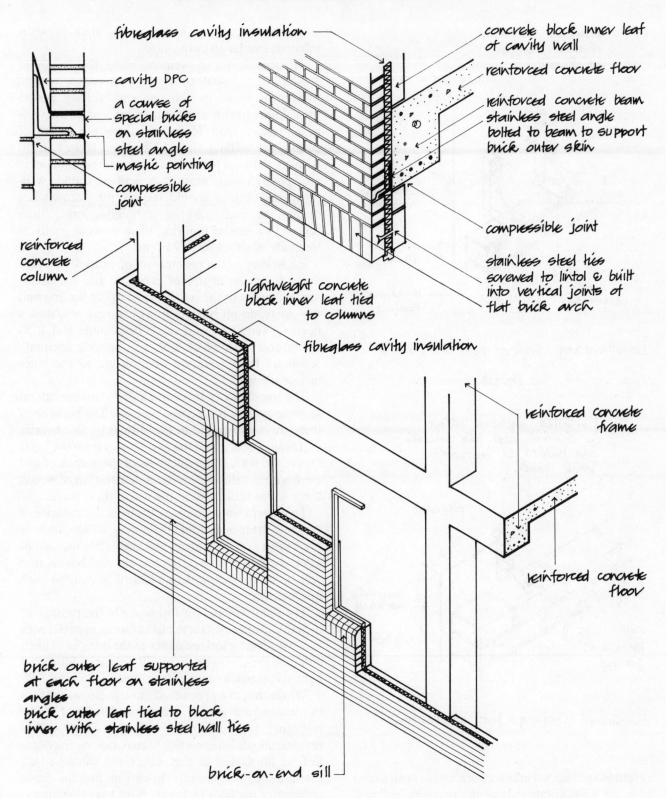

fibreglass cavity insulation

cavity DPC

a course of special bricks on stainless steel angle

mastic pointing

compressible joint

concrete block inner leaf of cavity wall

reinforced concrete floor

reinforced concrete beam stainless steel angle bolted to beam to support brick outer skin

compressible joint

stainless steel ties screwed to lintol & built into vertical joints of flat brick arch

reinforced concrete column

lightweight concrete block inner leaf tied to columns

fibreglass cavity insulation

reinforced concrete frame

reinforced concrete floor

brick outer leaf supported at each floor on stainless angles
brick outer leaf tied to block inner with stainless steel wall ties

brick-on-end sill

Brick cladding to reinforced concrete structural frame

Fig. 127

129

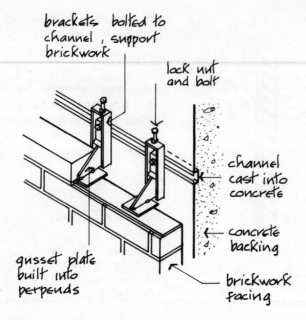

brackets bolted to channel, support brickwork

lock nut and bolt

channel cast into concrete

concrete backing

gusset plate built into perpends

brickwork facing

Loadbearing fixing for brickwork

Fig. 128

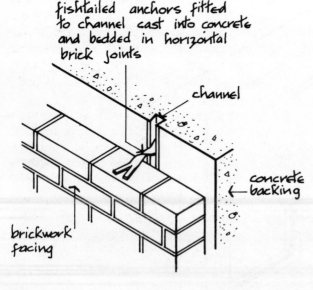

fishtailed anchors fitted to channel cast into concrete and bedded in horizontal brick joints

channel

concrete backing

brickwork facing

Restraint fixing for brickwork

Fig. 129

columns and bedded in brickwork and to both leaves of cavity walls at intervals similar to cavity wall ties.

Where cavity walling is built up to the face of columns of the structural frame and supported at every third storey, support and restraint against lateral forces is provided by anchors fitted to cast-in

channels and bedded in horizontal brick joints at intervals similar to cavity ties.

To provide for movement along the length of walling it is usual to form continuous vertical movement joints to coincide with vertical movement joints in the structural frame and at intervals of not more than 15 metres along the length of continuous walling and at 7.5 metres from bonded corners, with the joints filled with compressible strip and pointed with mastic. A wall of sound, well-burnt clay bricks should require no maintenance during the useful life of a building other than renewal of mastic pointing of movement joints at intervals of about 20 to 25 years.

Resistance to the penetration of wind-driven rain depends on the degree of exposure and the necessary thickness of the outer leaf of cavity walling and the cavity. In all but the most exposed positions a normal cavity wall of a single brick outer leaf, a 50 cavity and inner block leaf, will provide adequate resistance to penetration of moisture to the inner face of the wall.

The use of cavity trays and a dpc at all horizontal stops to cavities is accepted practice. The purpose of these trays, illustrated in Fig. 126, is to direct water that may collect inside the cavity away from the inner face of the wall. If the thickness of the outer leaf and the cavity is sufficient to resist penetration of water, there seems little logic in the use of these trays.

To prevent water soluble salts from the concrete of concrete frames finding their way to the face of brickwork and so causing unsightly efflorescence of salts, the face of concrete columns and beams that will be in contact with brickwork is painted with bitumen.

The requirements for resistance to the passage of heat usually necessitate the use of some material with comparatively good resistance to the transfer of heat, either in the cavity as cavity fill or partial fill with a lightweight block inner leaf, as illustrated in Fig. 127.

Where the cavity runs continuously across the face of the structural frame, as illustrated in Fig. 127, the resistance to the transfer of heat of the wall is uninterrupted. Where a floor slab supports the outer leaf, as illustrated in Fig. 126, there will be to an extent a cold bridge as the brick slips and the dense concrete of the floor slab will afford less resistance to the transfer of heat than the main cavity wall. The very small area of floor and ceiling that may well be colder is generally of little consequence and unlikely to encourage damp staining.

130

FACINGS APPLIED TO SOLID AND CAVITY WALL BACKING

The word facings is used to describe comparatively thin, non-structural slabs of natural or reconstructed stone, faience, ceramic and glass tiles or mosaic which are fixed to the face of and supported by solid background walls or to structural frames as a decorative finish. Common to the use of these non-structural facings is the need for the background wall or frame to support the whole of the weight of the facing at each storey height of the building or at vertical intervals of about three metres, by means of angles or dowels. In addition to the support fixings, restraint fixings are necessary to locate the facing units in true alignment and to resist wind pressure and suction forces.

To allow for elastic and flexural movements of the structural frame and differential thermal and moisture movements, there must be flexible horizontal joints below support fixings and vertical movement joints at intervals along the length of the facings. Both horizontal and vertical movement joints must be sufficiently flexible to accommodate anticipated movements and be water resistant to prevent penetration of rainwater.

Natural and reconstructed stone facings

Natural and reconstructed stone facings are applied to the face of buildings to provide a decorative finish to simulate the effect of solidity and permanence traditionally associated with solid masonry. Because of the very considerable cost of preparation and fixing, this type of facing is mostly used for prestige buildings such as banks and offices in city centres.

Granite is the natural stone much favoured for use as facing slabs for the hard, durable finish provided by polished granite and the wide range of colours and textures available from both native and imported stone. Polished granite slabs are used for the fine gloss surface that is maintained throughout the useful life of a building. To provide a more rugged appearance the surface of granite may be honed to provide a semi-polish, flame textured to provide random pitting of the surface or surface tooled to provide a more regular rough finish. Granite facing slabs are generally 40 thick for work more than 3.7 above ground and 30 thick for work less than 3.7 above ground.

Limestone, such as the native Portland, Bath or Clipsham, is used as facings, by and large, to resemble solid ashlar masonry work, the slabs having a smooth finish to reveal the grain and texture of the material. These comparatively soft limestones suffer a gradual change of colour over the course of years and this weathering is said to be an attractive feature of these limestones. Limestone facing slabs are 75 thick for work more than 3.7 above ground and 50 thick for work less than 3.7 above ground.

Hard limestones, including Roman stone and a number of very dense stones from France and Germany, are much used as facings for the hardness and durability of the materials. This type of stone is generally used as flat, level finished, facing slabs in thicknesses of 40 for work more than 3.7 above ground and 30 for work less than 3.7 above ground.

Sandstones, such as York stone, Darley Dale, Blaxter and Hollington, are used as facing slabs. Some care and experience are necessary in the selection of these native sandstones as the quality, and therefore the durability, of the stone may vary between stones taken from the same quarry. This type of stone is chosen for the colour and grain of the natural material whose colour will gradually change over some years of exposure. Because of the coarse grain of the material it may stain due to irregular run-off of water down the face. As with limestone, sandstone facing slabs are usually 75 thick for work 3.7 above ground and 50 thick for work less than 3.7 above ground.

Marble is less used for external facings in northern European climates, as polished marble finishes soon lose their shine. Coarser surfaces, such as honed or eggshell finishes, will generally maintain their finish, providing white or travertine marble is used. Marble facing slabs are 40 thick for work 3.7 above ground and 30 thick for work below that level. Reconstructed stone made with an aggregate of crushed natural stone is used as facing slabs as if it were the natural material, in thicknesses the same as that for the natural stone.

Fixing natural and reconstructed stone facings

The size of natural and reconstructed stone facing slabs is generally limited to about 1.5 in any one or both of the face dimensions or such less size as is practical to win from the quarry. Facing slabs are fixed so that there is a cavity between the back of the slabs and the background wall or frame to allow for fixings, tolerances in sawn thickness of slabs and to provide a cavity to minimise penetration of water to the backing. Where water might be trapped in the cavity behind the facing slabs on cavity trays weep holes should be provided.

The type of fixings used to support and secure facing slabs in position are:

- Loadbearing fixings
- Restraint fixings
- Combined loadbearing and restraint fixings

- Face fixings
- Soffit fixings

Loadbearing fixings

These support fixings are made from one of the corrosion resistant metals such as stainless steel, aluminium bronze or phosphor bronze. Stainless steel is the general description for a group of steel

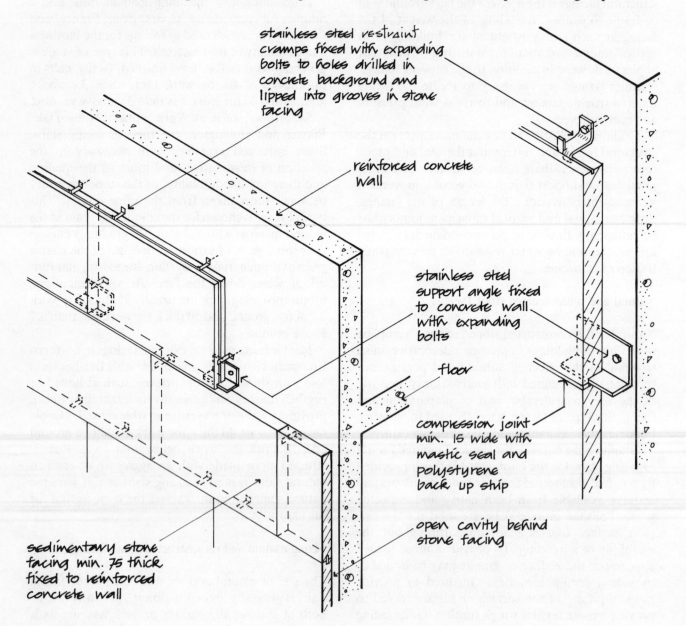

stainless steel restraint cramps fixed with expanding bolts to holes drilled in concrete background and lipped into grooves in stone facing

reinforced concrete wall

stainless steel support angle fixed to concrete wall with expanding bolts

floor

compression joint min. 15 wide with mastic seal and polystyrene back up strip

open cavity behind stone facing

sedimentary stone facing min. 75 thick fixed to reinforced concrete wall

Stone facing to solid background concrete wall

Fig. 130

132

alloys containing chromium and other elements. The type of stainless steel commonly used for structural fixings is austenitic stainless steel.

Loadbearing fixings take the form of either stainless steel angle supports to the lower edge of each slab or stainless steel corbel plates set in slots cut in the back of stones as illustrated in Figs 130 and 131. The size of the angle or corbel plate supports depends on the weight of the slabs to be supported. Common practice is to support each facing slab on two supports with the angle or corbel supports fixed

centrally on vertical joints between slabs so that each supports two slabs. Angle or corbel plate supports should be at least 75 wide. At vertical movement joints two supports are used, one each side of the joint, to the lower edge or lower part of the two stone slabs each side of the joint. These supports should be at least 50 wide. The usual method of fixing angle supports to the supporting background is with expanding bolts fitted to holes drilled in the concrete or solid brick backing as illustrated in Fig. 132. Where the background is sound and solid this form of fixing, made by the stone fixer, can be accurately located whereas preformed pockets or holes in the background may not coincide with the required fixing position. As an alternative bolts may be fitted to a metal channel cast into concrete as illustrated in Fig. 133. Where support fixings are to be built into a brick or block wall backing it is usual to build the background wall at the same time that the stone slabs are fixed, so that the stone slabs may be set in position and fishtailed ended corbel plates can be solidly bedded in brick or block horizontal courses two-thirds of the thickness of the background wall or not less than 100.

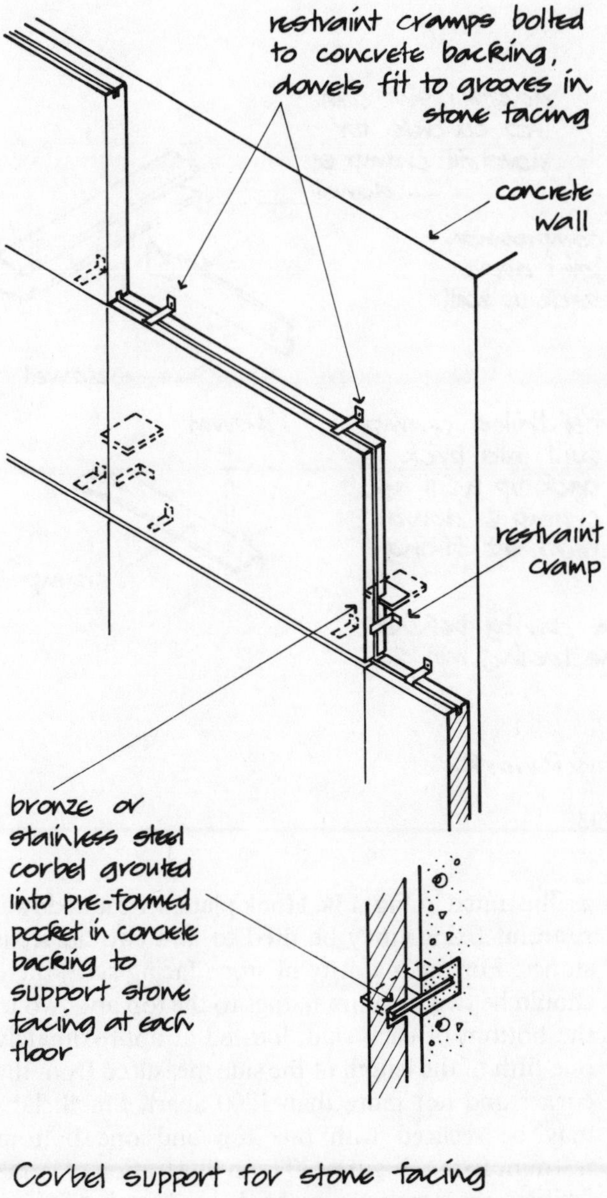

restraint cramps bolted to concrete backing, dowels fit to grooves in stone facing

concrete wall

restraint cramp

bronze or stainless steel corbel grouted into pre-formed pocket in concrete backing to support stone facing at each floor

Corbel support for stone facing

Fig. 131

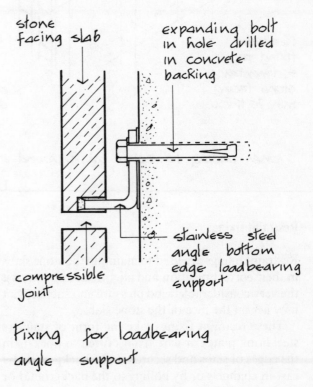

stone facing slab

expanding bolt in hole drilled in concrete backing

compressible joint

stainless steel angle bottom edge loadbearing support

Fixing for loadbearing angle support

Fig. 132

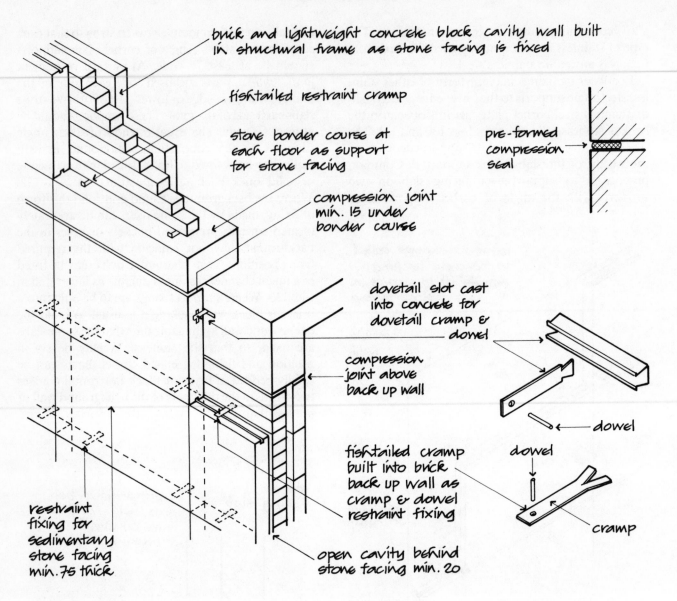

brick and lightweight concrete block cavity wall built in structural frame as stone facing is fixed

fishtailed restraint cramp

stone border course at each floor as support for stone facing

compression joint min. 15 under border course

pre-formed compression seal

dovetail slot cast into concrete for dovetail cramp & dowel

compression joint above back up wall

fishtailed cramp built into brick back up wall as cramp & dowel restraint fixing

dowel

dowel

cramp

open cavity behind stone facing min. 20

restraint fixing for sedimentary stone facing min. 75 thick

Stone facing to solid background of brickwork

Fig. 133

Restraint fixings

Restraint fixings are used to maintain the stone slabs in their correct position and alignment and to resist the very considerable wind pressure and suction that may act on the face of the stone slabs.

These restraint fixings take the form of stainless steel hook plates or wire fixings fitted to pockets in the edges of slabs and secured to the background by cast-in channels or by bolting to the background or by dowels fitted to plates secured in cast-in channels as illustrated in Fig. 134. Hook plate, wire and dowel restraint fixings may be used to join two adjacent stones. For the majority of stone facing slabs there should be two restraint fixings to the top and two to the bottom of each slab, located at approximately one-fifth of the length of the side measured from the corner and not more than 1200 apart. Small slabs may be secured with one top and one bottom restraint fixing. Figure 130 is an illustration of stone facing slabs supported by angle loadbearing fixings and restrained by hooked plate restraint fixings.

plate is inclined upwards at 15° and thus acts as restraint.

Face fixings

As an alternative to support and restraint fixing of stone facing slabs, face fixing may be used for thin slabs. Each stone facing slab is drilled for and fixed to a solid background with at least four stainless steel or non-ferrous bolts. The stone slabs are bedded in position on dabs of lime putty or weak mortar which is spread on the back of the slabs and the slabs are secured with expanding bolts and washers as illustrated in Fig. 135. The holes drilled in the stone for the bolts are then filled with pellets of stone to match the stone of the slabs. Joints between stones are filled with gunned-in mastic sealant to provide a weathertight joint and to accommodate differential thermal and structural movement. The whole of the weight of the stone slabs is supported by the bolts that must have a sufficient section to support the weight in shear and the bolts must be accurately set in place and strongly secured to the solid backing if the not uncommon failure of this method of fixing is to be avoided.

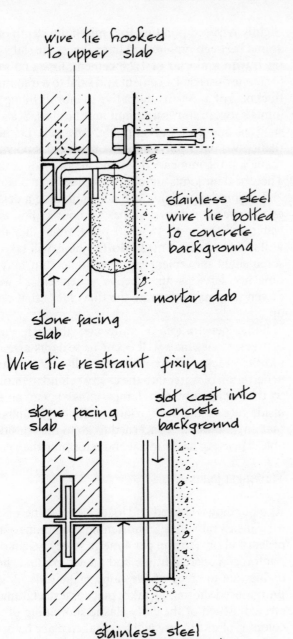

wire tie hooked to upper slab

stainless steel wire tie bolted to concrete background

mortar dab

stone facing slab

Wire tie restraint fixing

stone facing slab

slot cast into concrete background

stainless steel cramp and dowels fitted to cast in slot

Cramp and slot restraint fixing

Fig. 134

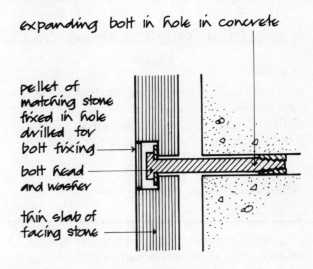

expanding bolt in hole in concrete

pellet of matching stone fixed in hole drilled for bolt fixing

bolt head and washer

thin slab of facing stone

Face fixing for stone cladding

Fig. 135

Combined loadbearing and restraint fixings

Loadbearing fixings may be designed and used as restraint fixings, as illustrated in Fig. 133, where angle fixings have a loose or welded on dowel pin to act as restraint or the protruding flange of a corbel

Soffit fixings

Support and fixing of stone facings to soffits is effected by the use of hangers and plates, cramps or dowels that are suspended in slot hangers cast in the

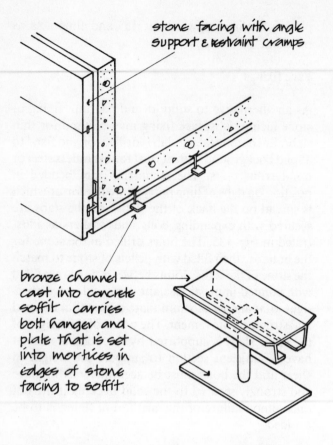

stone facing with angle support & restraint cramps

bronze channel cast into concrete soffit carries bolt hanger and plate that is set into mortices in edges of stone facing to soffit

Soffit fixing for stone facing

Fig. 136

structural soffit. The system illustrated in Fig. 136 consists of slot anchors that are cast into the concrete soffit to support hangers and plates that fit to slots in joints between stones to take the whole of the weight of the stone lining to the soffit.

Joints between stone slabs

Joints between stone facing slabs should be sealed as a barrier to the penetration of rainwater running off the face of the slabs. Where rainwater penetrates the joints between stone slabs it will be trapped in the cavity between the slabs and the background wall and will not evaporate to air during dry spells and may cause conditions of persistent damp. Open or butt joints between slabs should be avoided in external face work.

The joints between sedimentary stone slabs, such as limestone and sandstone, may be filled with a mortar of cement, lime and sand (or crushed natural stone) mix 1:1:6 and finished with either flush or

slightly recessed pointing to a minimum depth of 5. Joints between granite and hard limestone slabs are filled with a mortar of 1:2:8 cement, lime and sand (or stone dust) or 4:1 cement and sand to a minimum thickness of 3. As an alternative to mortar filling the joints between stones a sealant may be used. Sealants such as one part polysulphide, one part polyurethane, two parts polysulphide or two parts polyurethane are recommended for the majority of stones. These sealant joints should be not less than 5 wide.

The jointing sealants will accommodate a degree of movement between stones without failing as a water seal for up to 15 to 20 years, when they may well need to be reformed. Mortar joints will take up some slight movement between stones but may in time not serve as an effective water seal as wind driven rain may penetrate the fine cracks that open up.

Some penetration of rainwater through joints between stones may well occur as sealants age and mortar cracks. There will generally be no large penetration of water into the cavity behind the stones so that unless obvious damp stains appear on the inside face of the background walling it is unlikely that anyone will be concerned to go to the considerable expense of hacking out and reforming the joints.

Movement joints

As a building is erected and loads increase there is an early measurable elastic shortening of columns of the order of about 2.5 mm per 4 metre storey height of a reinforced concrete frame and a later gradual shortening due to creep of the same magnitude. A less pronounced shortening takes place as a steel frame is erected. Much of the early elastic shortening of the columns of a structure will have taken place before a wall cladding is fixed. The long-term shortening of reinforced concrete columns, through creep, has to be allowed for in horizontal movement joints. Differential temperature and moisture movements of a wall facing relative to the supporting structure will generally dictate the need to allow some movement of joints and fixings. There will be, for example, very considerable temperature differences between facing slabs on an exposed south facing wall and the structure behind so that differential thermal movement has to be allowed for both in joints and support and restraint fixings.

A general recommendation in the fixing of stone facing slabs is that there should be horizontal movement joints at each storey height below load-bearing support fixings or not more than 3 metres. These joints are usually 10 to 15 deep and filled with one of the elastic sealants. Where so wide a joint would not be acceptable in facework finished with narrow joints it is usual to accommodate movement in narrower sealant filled horizontal joints to all the facework. Vertical movement joints are formed in facework where these joints occur in the structure to allow for longitudinal structural, thermal and moisture movements. A continuous vertical joint is formed between stone facings and filled with sealant.

Faience slabwork

Faience is the term used to describe fire glazed stoneware in the form of slabs that are used as a facing to a solid background wall. The best quality slabs are made from stoneware which shrinks and deforms less on firing than does earthenware. The fired slab is glazed and then refired to produce a fire glazed finish. The slabs are usually 300 × 200, 450 × 300 or 610 × 450 and 25 to 32 thick. They form a durable, decorative facing to solid walls. The glazed finish, which will retain its lustre and colour indefinitely, needs periodic cleaning, especially in polluted atmospheres.

Faience slabwork was much used as a facing in the 1930s in this country, as a facing to large buildings such as cinemas. This excellent facing material has since then lost favour. When first used the slabs were fixed with cement mortar dabs to a keyed brick background. This unsatisfactory method of fixing made no allowance for differential movements and has been abandoned in favour of support fixings to each slab and restraint fixings and movement joints in the same way that stone facings are fixed.

Terra cotta

Terra cotta (burnt earth) was much used in Victorian buildings as a facing because it is less affected by polluted atmospheres than natural limestone and sandstone facings. Fired blocks of terra cotta, with a semi-glaze self finish, were moulded in the form of natural stone blocks to replicate the form and detail of the stonework buildings of the time. The plain and ornamental blocks were made hollow to reduce and control shrinkage of the clay during firing. In use the hollows in the blocks were filled with concrete and the blocks were then laid as if they were natural stone. Well burned blocks of terra cotta are durable even in heavily polluted atmospheres. This labour intensive system of facing is little used today.

Tiles and mosaic

Tile is the term used to describe comparatively thin, small slabs of burnt clay or cast concrete up to about 300 square and 12 thick. These small units of fired clay and cast concrete are used as a facing to structural frames and solid background walls. For many years practice has been to bond tiles directly to frames and walls with cement mortar dabs which by themselves provide sufficient adhesion to maintain individual tiles in place. This system of adhesion does not make any allowance for differential movements between the frame, background walls and tiles, other than in the joints between tiles, which can be considerable, particularly with in situ cast concrete work. To make allowance for movements in the structure and the facing, tiles should be supported and restrained by cramps that provide a degree of flexibility between the facing and the background. Plainly it would be both tedious and expensive to fix individual tiles in this way. For economy and ease of fixing, the tiles can be cast on to a slab of plain or reinforced concrete which is then fixed in the same way as stone facing slabs.

Mosaic is the term used to describe small squares of natural stone, tile or glass set out in some decorative pattern. The units of mosaic are usually no larger than 25 square. A mosaic finish as an external facing should be used as a facing to a cast concrete slab in the same way as tiles.

CLADDING PANELS

Precast concrete cladding panels

The aesthetic, economic and constructional advantages of the use of precast concrete cladding slabs or units as a facing and walling to framed structures were demonstrated by Le Corbusier in the multi-storey housing development, the Unité at Marseilles, which was completed in 1952.

The use of precast concrete and the design of this one building had a profound influence on the use of precast concrete for many years. The advantages of repetitive casting, speed of erection largely independent of weather and the rugged appearance of the material that was a vogue of the 1950s and 1960s led to the extensive use of this system of wall cladding.

Precast concrete cladding panels or units are usually storey height as illustrated in Fig. 137 or column spacing wide as spandrel or undersill units for support and fixing to the structural frame. Precast concrete cladding units are hung on and attached to frames as a self-supporting facing and wall element which may combine all of the functional requirements of a wall element.

Precast concrete cladding units are cast with either the external face up or down in the moulds, depending on convenience in moulding and the type of finish. Where a finish of specially selected aggregate is to be exposed on the face, the face up method of casting is generally used for the convenience and accuracy in applying the special finish to the core concrete of the panel. Cladding units that are flat or profiled are generally cast face down for convenience in compacting concrete into the face of the mould bed.

Strongly constructed moulds of timber, steel or

glass fibre reinforced plastic are laid horizontal, the reinforcing cage and mesh is positioned in the mould and concrete is placed and compacted. For economy in the use of the comparatively expensive moulds it is essential that there be a limited number of sizes, shapes and finishes to cladding units to obtain the economic advantage of repetitive casting.

There is no theoretical limit to the size of precast units, providing they are sufficiently robust to be handled, lifted and fixed in place, other than limitations of the length of a unit that can be transported and lifted. In practice cladding units are usually storey height for convenience in transport and lifting and fixing in place. Cladding units two or more storeys in height have to be designed, hung and fixed to accommodate differential movements between the frame and the units, which are multiplied by the number of storeys they cover.

The initial wet plastic nature of concrete facilitates the casting of a wide variety of shapes and profiles from flat solid web enclosing panels to the comparatively slender solid sections of precast concrete frames for windows.

The limitation of width of cladding units is determined by facilities for casting and size for transport and lifting. The width of the units cast face up is limited by ease of access to placing the face material in the moulds. The usual width of storey height panels is from 1200 to 1500, or the width of one or two structural bays.

For strength and rigidity in handling, transport, lifting and support and fixing, and to resist lateral wind pressures, cladding units are reinforced with a mesh of reinforcement to the solid web of units and a cage of reinforcement to vertical stiffening ribs and horizontal support ribs. Figure 138 is an illustration of a storey height cladding unit.

The vertical stiffening ribs are designed for strength in resisting lateral wind pressures on the units between horizontal supports and strength in supporting the weight of the units that are either hung from or supported on the horizontal support ribs. The least thickness of concrete necessary for the web and the ribs is dictated largely by the cover of concrete necessary to protect reinforcement from corrosion, for which a minimum web thickness of 85 or 100 is usual. The necessary cover of concrete to reinforcement makes this system of walling heavy, cumbersome to handle and fix and bulky looking.

Storey height precast concrete cladding is supported by the structural frame, either by a horizontal

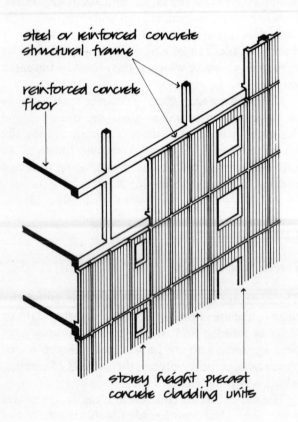

steel or reinforced concrete structural frame

reinforced concrete floor

storey height precast concrete cladding units

Storey height precast concrete cladding

Fig. 137

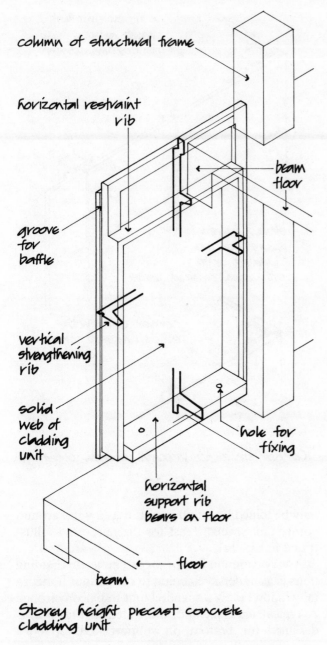

column of structural frame

horizontal restraint rib

beam floor

groove for baffle

vertical strengthening rib

solid web of cladding unit

hole for fixing

horizontal support rib bears on floor

floor

beam

Storey height precast concrete cladding unit

Fig. 138

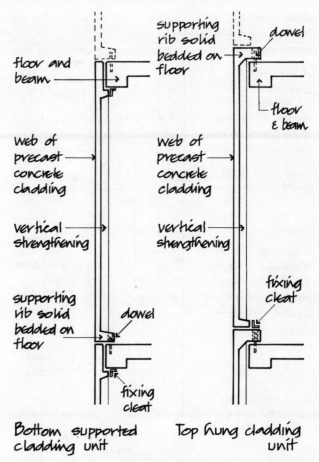

floor and beam

web of precast concrete cladding

vertical strengthening

supporting rib solid bedded on floor

dowel

fixing cleat

Bottom supported cladding unit

supporting rib solid bedded on floor

dowel

floor & beam

web of precast concrete cladding

vertical strengthening

fixing cleat

Top hung cladding unit

Fig. 139

support rib at the bottom of the units or hung on a horizontal support rib at the top of the units, as illustrated in Fig. 139. Bottom support is preferred as the concrete of the unit is in compression and less likely to develop visible cracks and crazing than it is when top hung. Whichever system of support is used, the horizontal support rib must have an adequate projection for bearing on structural floor slabs or beams and for the fixings used to secure the units to the frame. At least two mechanical support and two restraint fixings are used for each unit. The usual method of fixing at supports is by the use of steel or non-ferrous dowels that are grouted into 50 square pockets in the floor slab. The dowel is then grouted into a 50 diameter hole in the support rib, as illustrated in Fig. 140. The advantage of this dowel fixing is that it can readily be adjusted to inaccuracies in the structure and the panel. Dowel fixings serve to locate the units in position and act as restraint fixings against lateral wind pressures.

Restraint fixings to the upper or lower horizontal ribs of cladding units, depending on whether they are top or bottom supported, must restrain the unit in place against movements and lateral wind pressure. The restraint fixing most used is a non-ferrous or stainless steel angle cleat that is either fixed to a slotted channel cast in the soffit of beams or slabs or more usually by expanding bolts fitted to holes drilled in the concrete. The cleat is bolted to a cast-in stud protruding from the horizontal rib of the unit as illustrated in Fig. 140. The slotted hole in the

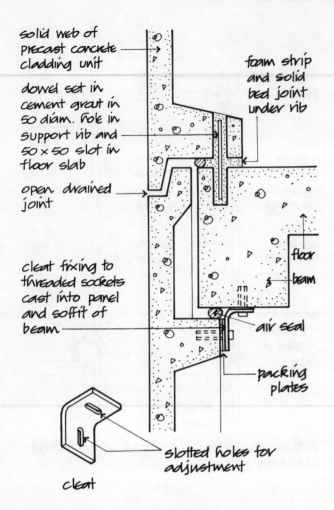

solid web of precast concrete cladding unit

dowel set in cement grout in 50 diam. hole in support rib and 50 x 50 slot in floor slab

open drained joint

cleat fixing to threaded sockets cast into panel and soffit of beam

foam strip and solid bed joint under rib

floor

beam

air seal

packing plates

cleat

slotted holes for adjustment

Fixing for bottom supported precast concrete cladding units

Fig. 140

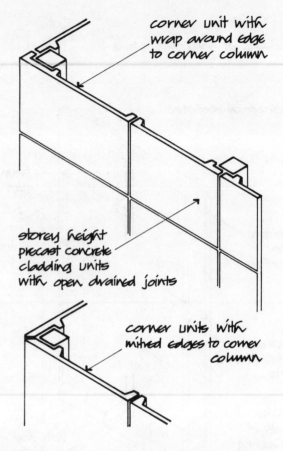

corner unit with wrap around edge to corner column

storey height precast concrete cladding units with open drained joints

corner units with mitred edges to corner column

Corner units to precast concrete cladding

Fig. 141

downstand flange of the cleat allows some vertical movement between the frame and the cladding.

Another system of fixing combines support fixing by dowels with restraint fixing by non-ferrous flexible straps that are cast into the units and fit over the dowel fixing. Support and restraint fixing may be provided by casting loops or hooked ends of reinforcement, protruding from the back of cladding units, into a small part of or the whole of an in situ cast concrete member of the structural frame. The disadvantage of this method is the site labour required in making a satisfactory joint and the rigidity of the fixing that makes no allowance for differential movements between structure and cladding.

At external angles on elevations cladding units may be joined by a mitre joint or as a wrap around corner unit specially cast for the purpose, as illustrated in Fig. 141.

A very common use for precast concrete cladding units is as undersill cladding to continuous horizontal windows or as a spandrel unit to balcony fronts. A typical undersill unit, illustrated in Fig. 142, is designed for bottom rib support and top edge restraint at columns.

The web, horizontal support and restraint ribs and the vertical stiffening ribs are similar in construction to storey height panels. Support and restraint fixing is through the bottom horizontal rib bearing on the floor slab with dowel fixing and restraint fixing by cleats fixed to columns and ribs, as illustrated in Fig. 143. As an alternative to bottom support these units can be top supported on an in situ cast concrete beam at sill level so that several undersill cladding units may be used between widely spaced columns. Joints are made as either sealed or open drained joints, similar to those for storey height panels.

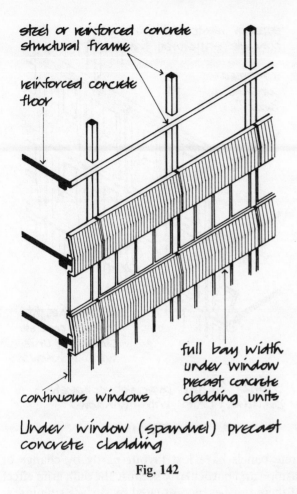

steel or reinforced concrete
structural frame

reinforced concrete
floor

full bay width
under window
precast concrete
cladding units

continuous windows

Under window (spandrel) precast
concrete cladding

Fig. 142

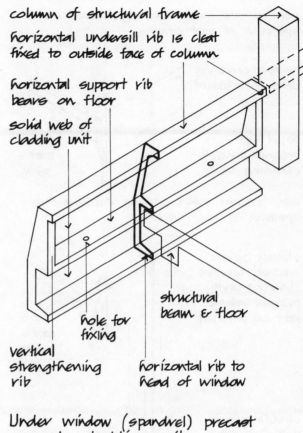

column of structural frame

horizontal undersill rib is cleat
fixed to outside face of column

horizontal support rib
bears on floor

solid web of
cladding unit

structural
beam & floor

hole for
fixing

vertical
strengthening
rib

horizontal rib to
head of window

Under window (spandrel) precast
concrete cladding unit

Fig. 143

At parapet level of buildings faced with precast concrete cladding panels either a special cladding unit is used or a special parapet unit is cast, as illustrated in Fig. 144.

Because of the plastic nature of wet concrete it is possible to cast cladding units in a variety of profiled and textured finishes and to include openings for windows in individual cladding units, within the need for repetitive casting for economy. The limitations to the complexity and fineness of detail of the textures and profiles that can be achieved are the need for reinforcement and cover of concrete for stiffening ribs at edges and around openings and the size of the aggregate in concrete that limits fine detail. Figure 145 is an illustration of a profiled window unit.

Surface finishes

Due to compaction of wet concrete in the mould, the lower face of the concrete consists of a water rich mix of the fine particles of cement and aggregate. On drying this thin layer of cement rich material shrinks and forms a surface of irregular fine cracks and the surface may show marked colour differences due to variations in placing, compacting and mixing of the concrete. Because such a surface is not generally an acceptable finish the exposed faces of precast concrete are usually treated to reveal the underlying aggregate.

To provide an acceptable finish to the exposed faces of precast concrete panels it is practice to provide what is sometimes called an 'indirect finish' by abrasive blasting, surface grinding, acid washing or tooling to remove the fine surface layer and expose the aggregate and cement below. This surface treatment has the general effect of exposing a surface of reasonably uniform colour and texture. This form of surface treatment can produce a fine smooth finish by light abrasive sand or grit blasting or grinding or a more coarse texture by heavy surface treatment.

It was the fashion for some years to use coarse finishes to precast concrete panels by exposing the surface aggregate by heavy abrasive blasting or tooling the surface to produce bush hammered,

141

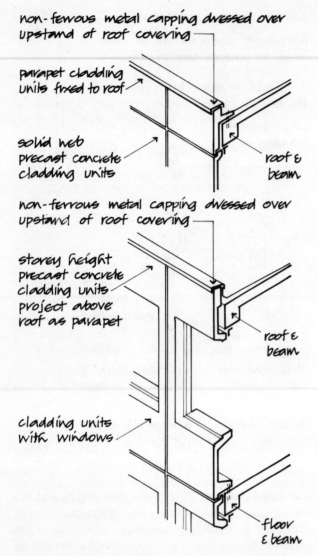

non-ferrous metal capping dressed over upstand of roof covering

parapet cladding units fixed to roof

solid web precast concrete cladding units

roof & beam

non-ferrous metal capping dressed over upstand of roof covering

storey height precast concrete cladding units project above roof as parapet

roof & beam

cladding units with windows

floor & beam

Parapets to precast concrete cladding units

Fig. 144

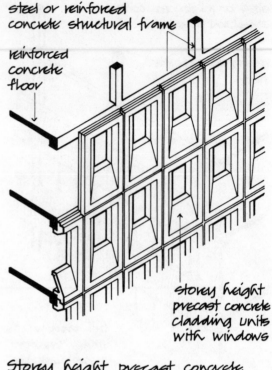

steel or reinforced concrete structural frame

reinforced concrete floor

storey height precast concrete cladding units with windows

Storey height precast concrete cladding units with windows

Fig. 145

chisel or pointed tool finishes to emphasise the rugged, heavy nature of the panels.

A variety of profiled finishes is produced by casting the panels face down in moulds against timber, etched metal or glass fibre formers to produce a distinct profiled finished face. The finished face of the panels is acid washed or abrasive blasted to remove the surface of fine particles of cement and expose the aggregate. Much favoured at one time was a board marked finish produced by casting on to the surface of boards which had been grit blasted so that the finished concrete surface displayed the grain of the wood.

Of recent years profiled finishes to precast concrete panels have lost favour partly by change of fashion and particularly because the dull, grim effect of these finishes, accentuated by surface staining, is not particularly attractive in the dull light of northern climates.

Exposed aggregate finishes are produced by casting the panels face up so that a selected aggregate of comparatively large stones may be spread over the wet compacted concrete and lightly compacted in place so that the aggregate is exposed. Once the concrete has hardened the surface is washed or blasted to remove traces of cement to expose the colour and texture of the aggregate. This form of rugged finish, which at one time was highly fashionable, has since lost favour.

Applied finishes

It is common today, where precast concrete cladding panels are used, to provide facing materials of brick or stone to the panels for the advantage of casting and fixing such traditional finishes to precast panels to minimise site labour.

Any sound, well-burnt type of brick that is reasonably frost resistant may be used as a facing

142

material and fixed to the precast concrete panels using the mould to produce the full range of brick construction features such as corbels, string courses, piers and arches.

The facing brickwork is bonded to the concrete panel either with a mechanical key or by stainless steel or nylon filament ties. A mechanical key can be provided where bricks with holes in them are used and cut along the length of the brick, so that the resulting semicircular grooves may provide a bond to the concrete. To ensure a good bond to the bricks it is essential to thoroughly saturate the bricks before the backing concrete is placed. Where ties are used to retain the face bricks in place the nylon filament, stainless steel wire or bars are threaded through holes in the brickwork and turned up between bricks as loops to bond with the backing concrete.

The face brickwork may be pre-pointed or post-pointed. For pre-pointing, fillets are suspended between the joints in brickwork which is laid in the bed of the mould and the pointing material of cement, lime and sand is then pumped into the space below the fillets. The face of the brickwork is protected from mortar by a cloth or paper impregnated with a retarder laid in the bed of the mould. The fillets are removed and the backing concrete placed and compacted over the bricks and around the tie loops. The face pointing, which is usually up to 25 deep, is allowed to harden before the concrete is placed. For post-pointing the bricks are laid in the bed of the mould between neoprene strips to provide the necessary recess of about 20 for pointing and the concrete backing is then cast on the bricks and consolidated. Pointing is carried out on site when the cast panels are in place.

This is a labour intensive way of providing a brick facing to concrete panels that at best will provide a simulation of traditional brickwork. Because there have to be both horizontal and vertical movement joints between the concrete panels, which will interrupt any attempt to copy loadbearing brickwork, it provides the possibility of setting the bricks in any pattern, other than the traditional bonded horizontal bonded pattern, for purely decorative purposes.

Natural and reconstructed stone facing slabs are used as a decorative finish to precast concrete panels. Any of the natural or reconstructed stones used for stone facework to solid backgrounds may be used for facings to precast concrete panels. For ease of placing the stone facing slabs in the bed of the mould, it is usual to limit the size of the panels to not more than 1.5 metres in any one dimension. Granite and hard limestone slabs not less than 30 thick and limestone, sandstone and reconstructed stone slabs not less than 50 thick are used.

The facing slabs are secured to and supported by the precast panel through stainless steel corbel dowels at least 4.7 mm in diameter, that are set into holes in the back of the slabs and cast into the concrete panels at the rate of at least 11 per metre square of panel and inclined at 45° or 60° to the face of the panel. Normal practice is that about half of the dowels are inclined up and half down, relative to the vertical position of the slab when in position on site. The dowels are set in epoxy resin in holes drilled in the back of the slabs. Flexible grommets are fitted around the dowels where they protrude from the back of the slab. These grommets, which are cast into the concrete of the panel, together with the epoxy resin bond of the dowel in the stone slab, provide a degree of flexibility to accommodate thermal and moisture movement of the slab relative to that of the supporting precast concrete cladding panel. All joints between the stone facing slabs are packed with closed cell foam backing or dry sand and all joints in the back of the stone slabs are sealed with plastic tape to prevent cement grout running in. When the precast panel is taken from the mould the jointing material is removed for mortar or sealant jointing.

To prevent the concrete of the precast panel bonding to the back of the stone slabs either polythene sheeting or a brushed on coating of clear silicone waterproofing liquid is applied to the whole of the back of the slabs. The purpose of this debonding layer is to allow the facing slabs free movement relative to the precast panel due to differential movements of the facing and the backing.

The necessary joints between precast concrete cladding panels faced with stone facing slabs are usually sealed with a sealant to match those between the facing slabs.

Joints between precast concrete cladding panels

The joints between cladding panels must be sufficiently wide to allow for inaccuracies in both the structural frame and the cladding units, to allow unrestrained movements due to shortening of the frame and thermal and moisture movements and at the same time exclude rain.

The two systems of making joints between units

are the face sealed joint and the open drained and rebated joint.

Sealed joints are made watertight with a sealant that is formed inside the joint over a backing strip of closed cell polyethylene, at or close to the face of the units as illustrated in Fig. 146. The purpose of the backing strip is to ensure a correct depth of sealant. Too great a depth or width of sealant will cause the plastic material of the sealant to move gradually out of the joint, due to its own weight. Sealant material is applied by a gun and compacted and shaped by hand. The disadvantage of sealant joints is that there is a limitation to the width of joint in which the sealant material can successfully be retained and that the useful life of the material is from 15 to 20 years, as it oxidises and hardens with exposure to sunlight and has to be raked out and renewed. Sealed joints are used in the main for the smaller cladding units.

The sealants most used for joints between precast concrete cladding panels are two part polysulphide, one part polyurethane, epoxy modified two part

polyurethane and low modulus silicone. Which of these sealants is used depends to an extent on experience in the use of a particular material and ease of application on site. The two part sealants require more skill in mixing the two components to make a successful seal than the one part material which is generally reflected in the relative cost of the sealants.

A closed cell polyethylene backing strip is rammed into the joint and the sealant applied by power or hand pump gun and compacted and levelled with a jointing tool.

Open drained joints between precast concrete cladding panels are more laborious to form than sealed joints and are mostly used for the larger precast panels where the width of the joint may be too wide to seal and where the visible open joint is

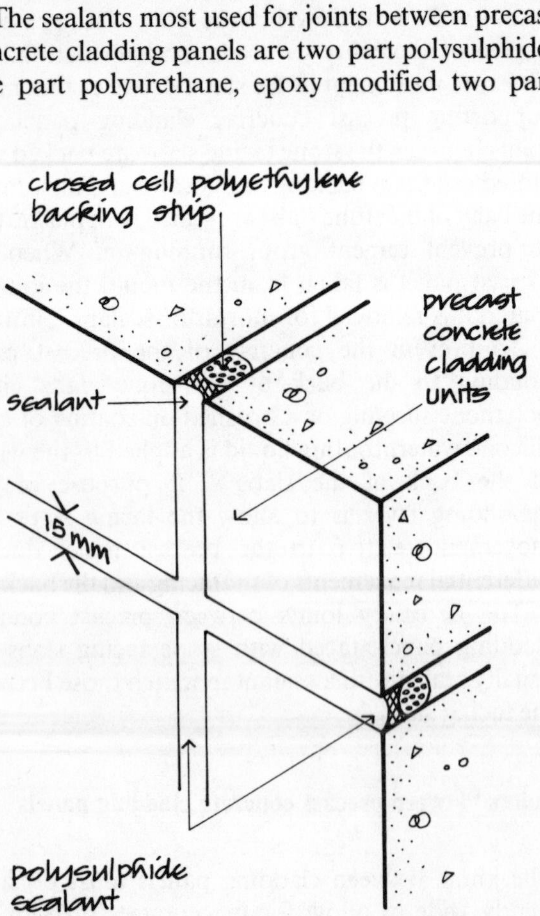

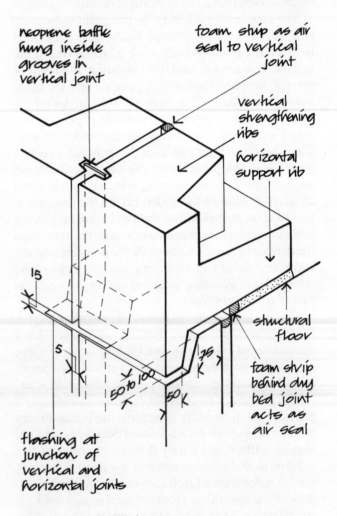

Sealed joints to precast concrete cladding units

Fig. 146

Open drained joints between storey height precast concrete cladding units

Fig. 147

used to emphasise the rugged, coarse textured finish to the panels.

Open joints are the most effective system of making allowance for inaccuracies and differential movements and serving as a bar to rain penetration without the use of joint filling material.

Horizontal joints are formed as open overlapping joints with a sufficiently deep rebate as a bar to rain penetration, as illustrated in Fig. 147. The rebate at the joint should be of sufficient section to avoid damage in transport, lifting and fixing in place. The thickness necessary for these rebates is provided by the depth of the horizontal ribs. The air seal formed at the back of horizontal joints is continuous in both horizontal and vertical joints as a seal against outside wind pressure and driving rain.

Vertical joints are designed as open drained joints in which a neoprene baffle is suspended inside grooves formed in the edges of adjacent units, as illustrated in Fig. 148. The open drained joint is designed to collect most of the rain in the outer zone of the joint in front of the baffle, which acts as a barrier to rain that may run or be forced into the joint by wind pressure. The baffle is hung in the joint so that to an extent there is a degree of air pressure equalisation each side of the baffle due to the air seal at the back of the joint. This air pressure equalisation acts as a check to wind driven rain that would otherwise be forced past the baffle if it were a close fit and there were no air seal at the back of the joint. At the base of each open drained joint there is a lead flashing, illustrated in Figs 147 and 148, that serves as a barrier to rain at the most vulnerable point of the intersection of horizontal and vertical joints. As cladding panels are fixed, the baffle in the upper joints is overlapped outside the baffle of the lower units.

Where there is a cavity between the back of the cladding units and an inner system of solid block walls or framing for insulation, air seals can be fitted between the frame and the cladding units.

It is accepted that the system of open joints between units is not a complete barrier to rain. The effectiveness of the joint depends on the degree of exposure to driving rain, the degree of accuracy in the manufacture and assembly of the system of walling and the surface finish of the cladding units. Smooth faced units will tend to encourage driven rain to sheet across and up the face of the units, and so cause a greater pressure of rain in joints than there would be with a coarse textured finished which will disperse

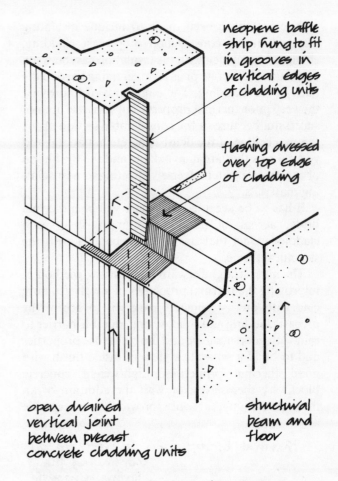

neoprene baffle strip hung to fit in grooves in vertical edges of cladding units

flashing dressed over top edge of cladding

open drained vertical joint between precast concrete cladding units

structural beam and floor

Open drained vertical joint between precast concrete cladding units.

Fig. 148

driven rain and wind and so reduce pressure on joints.

The backs of cladding panels will tend to collect moisture by possible penetration of rain through joints and from condensation of moisture laden air from outside and warm moist air from inside by vapour pressure, which will condense on the inner face of panels. Condensation can be reduced by the use of a moisture vapour check on the warm side of insulation as a protection against interstitial condensation in the insulation and as a check to warm moist air penetrating to the cold inner face of panels.

Precast concrete cladding panels are sometimes cast with narrow weepholes, from the top edge of the lower horizontal ribs out to the face, in the anticipation that condensate water from the back of the units will drain down and outside. The near certainty of these small holes becoming blocked by windblown debris makes their use questionable.

Attempts have been made to include insulating material in the construction of precast cladding, either as a sandwich with the insulation cast between two skins of concrete or as an inner lining fixed to the back of the cladding. These methods of improving the very poor thermal properties of concrete are not successful because of the considerable section of the thermal bridge of the dense concrete horizontal and vertical ribs that are unavoidable and the likelihood of condensate water adversely affecting some insulating materials.

It has to be accepted that there will be a thermal bridge across the horizontal support rib of each cladding panel that has to be in contact with the structural frame.

The most straightforward and effective method of improving the thermal properties of a wall structure clad with precast concrete panels is to accept the precast cladding as a solid, strong, durable barrier to rain with good acoustic and fire resistance properties and to build a separate system of inside finish with good thermal properties. Lightweight concrete blocks by themselves, or with the addition of an insulating lining, at once provide an acceptable

internal finish and thermal properties. Block wall inner linings should be constructed independent of the cladding panels and structural members, as far as practical, to reduce interruption of the inner lining as illustrated in Fig. 149.

GLASS FIBRE REINFORCED CEMENT CLADDING PANELS (GRC)

Glass fibre reinforced cement as a wall panel material was first used in the early 1970s after studies at the Building Research Establishment and the production of an alkali resisting glass fibre. The material has since been used as a lightweight substitute for precast concrete in wall cladding in the UK, in America, the Middle East and Japan.

The principal advantage of GRC as a wall panel material is weight saving as compared to similar precast concrete panels. Much of the mass of concrete used in panels is required as protection of the steel reinforcement against atmospheric chemical attack, whereas alkali resisting glass fibre, which is not subject to attack, can be used in panels with a skin thickness of 10 to 15 and a weight saving of about 80% of that of a comparable concrete panel. This weight reduction will afford substantial savings in transport, handling and erection costs and some small saving in structural frame members.

Because of the fine grain of the material that can be used in the manufacture of GRC and the freedom from the constraint of the need for steel reinforcement and its necessary cover against corrosion, this material can be formed in a wide variety of shapes, profiles and accurately finished mouldings. The material has inherently good durability and chemical resistance, is non-combustible, not susceptible to rot and will not corrode or rust stain. The limiting factors in the use of this material arise from relatively large thermal and moisture movements and the restricted ductility of the material.

The material is a composite of cement, sand and alkali resistant (AR) glass fibre in proportions of 40–60% cement, 20% water, up to 25% sand and 3.5–5% glass fibre by weight. The glass fibre is chopped to lengths of about 35 before mixing. It is formed in moulds by spray application of the wet mix, which is built up gradually to the required thickness and compacted by roller. After the initial 3 thickness has been built up it is compacted by roller to ensure a compact surface finish. For effective hand

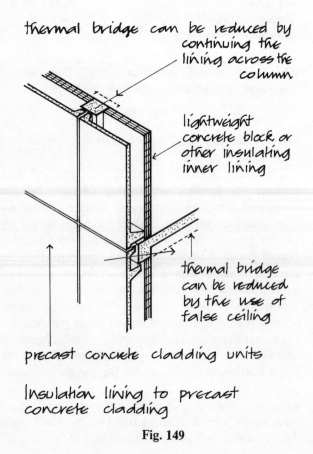

thermal bridge can be reduced by continuing the lining across the column

lightweight concrete block or other insulating inner lining

thermal bridge can be reduced by the use of false ceiling

precast concrete cladding units

Insulation lining to precast concrete cladding

Fig. 149

146

spraying the maximum width of panel is about 2 metres. For mass production runs of panel, a mechanised system is used with dual spray heads which spray fibre and cement, sand and water separately in the mould which moves under the fixed spray heads. The mechanised spray results in a greater consistency of the mix and a more uniform thickness of panel than is usually possible with hand spraying.

The moulds for GRC are either timber or the more durable GRP lined, timber framed types. Spray moulded GRC panels have developed sufficient strength 24 hours after moulding to be taken from moulds for curing.

The size of GRC cladding panels is limited by the method of production as to width and to the storey height length for strength, transport and lifting purposes. It is also limited by the very considerable moisture movement of the cement rich material, that may fail if moisture movement is restrained by fixings. The usual thickness of GRC single skin panels is 10 to 15.

As a consequence of moulding, the surface of a GRC panel is a cement rich layer which is liable to crazing due to drying shrinkage and to patchiness of the colour of the material due to curing. To remove the cement rich layer on the surface and provide a more uniform surface, texture and colour, the surface can be acid etched, grit blasted or smooth ground. Alternatively the panels can be formed in textured moulds so that the finished texture masks surface crazing and patchiness.

Using ordinary or rapid hardening Portland cement the natural colour of these panels is a light, dull grey. White or pigmented white cement can be used instead of Portland cement to produce a white or colour finish, which may well not be uniform, panel to panel. For a uniform colour finish that can be restored by repainting on site, coloured permeable coatings are used which have microscopic pores in their surface that allow a degree of penetration and evaporation of moisture that prevents blistering or flaking of the coating. Textured permeable finishes such as those used for external renderings and microporous matt and glass finish paints are used.

The thin single skin of GRC does not have sufficient strength or rigidity by itself to be used as a wall facing other than as a panel material of up to about 1 metre square, supported by a metal carrier system or bonded to an insulation core for larger panels, as illustrated in Fig. 150.

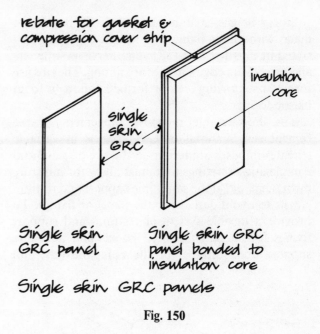

Single skin GRC panels

Fig. 150

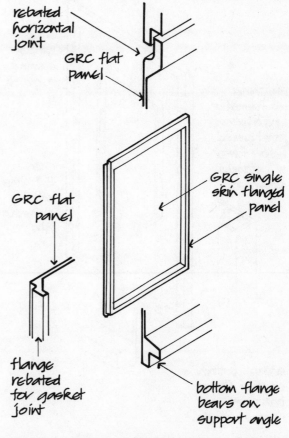

Single skin - flanged GRC panel

Fig. 151

Storey height, spandrel and undersill panels are made with either flanged or ribbed edges as illustrated in Figs 151 and 152, which serve as stiffeners and provide an edge surface for jointing. The ribs are formed by spraying over preformed hollow or foam plastic formers.

The considerable moisture movement of the cement rich material of GRC panels in variable climates imposes limitations to the size of panel and complexities in fixings that must allow for moisture movement and at the same time support and restrain panels to avoid damage to the panel or fixings. To provide a flexible system of restraint and support fixings for single skin GRC panels and adequate support and stiffness for the comparatively thin

material, a stud frame system is much used today for storey height panels. Stud frames are fabricated from cold formed steel sections with welded joints. The galvanised frame is attached to the back of the single skin panel during manufacture, with GRC strips that are rolled into the back of the panel over bent bars welded to the steel frame, as illustrated in Fig. 153. Two gravity anchors secure the GRC panel to the stud frame next to the two support fixings close to the lower edge of the panel. The bent bar restraint fixings to the back of the panel at once provide adequate fixing and accommodate moisture and thermal movements of the panel relative to the stud frame.

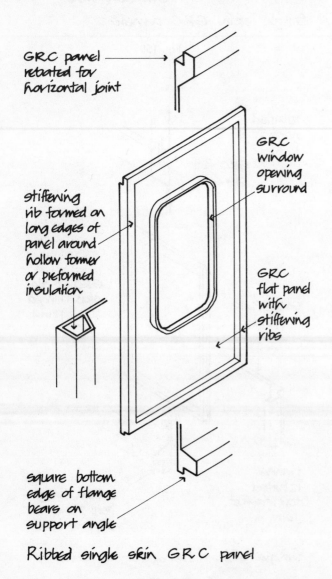

GRC panel rebated for horizontal joint

stiffening rib formed on long edges of panel around hollow former or preformed insulation

GRC window opening surround

GRC flat panel with stiffening ribs

square bottom edge of flange bears on support angle

Ribbed single skin GRC panel

Fig. 152

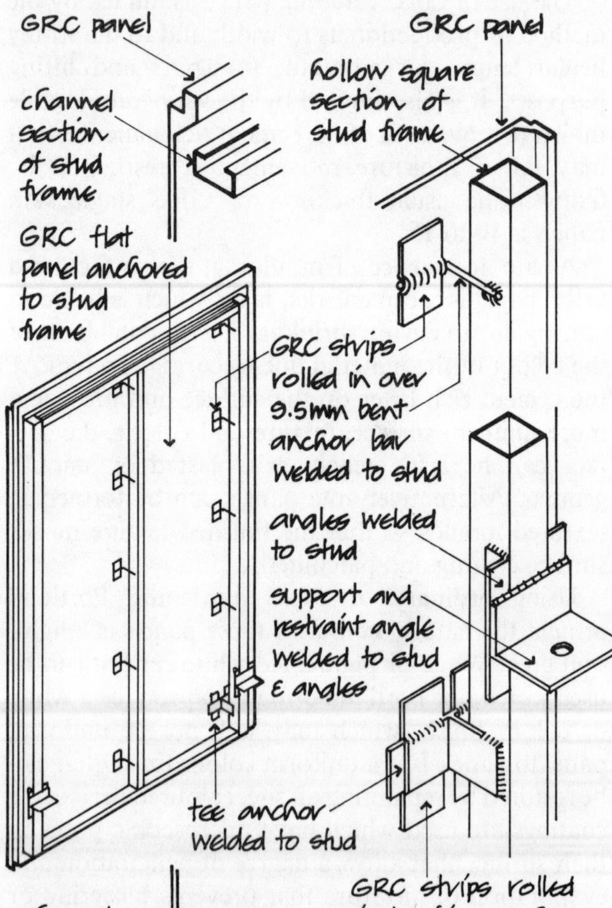

GRC panel

channel section of stud frame

GRC flat panel anchored to stud frame

GRC panel

hollow square section of stud frame

GRC strips rolled in over 9.5mm bent anchor bar welded to stud

angles welded to stud

support and restraint angle welded to stud & angles

tee anchor welded to stud

GRC strips rolled in to fix anchor to panel

GRC panel

channel section of stud frame

Stud frame support for single skin flanged GRC panel

Fig. 153

148

Figure 154 is an illustration of the fixings for a stud frame to a structural steel frame. This stud frame system of support and restraint for single skin panels which has been used for double storey height panels with success, is the preferred system of construction for GRC panels. A single skin of GRC has poor thermal properties, poor integrity against damage by fire and has to be stiffened for use as large panels.

Sandwich panels of GRC, made as a sandwich of two skins of GRC enclosing a core of insulation, have been used to combine the advantages of good thermal properties with the stiffness and strength gained from the sandwich construction. Figure 155 is an illustration of a typical sandwich panel of two skins of GRC formed around a core of insulation. Usual practice is to form ribs of GRC between the two skins across the core as stiffeners to the thin skins.

These composite panels, which in one unit fulfil the functional requirements of a wall, have to a large extent been abandoned in favour of stud frames. The disadvantages of a sandwich panel are the differences in both moisture and thermal movement between the inner and outer skins, which may at the least cause distortion of the panel face and at the most fracture

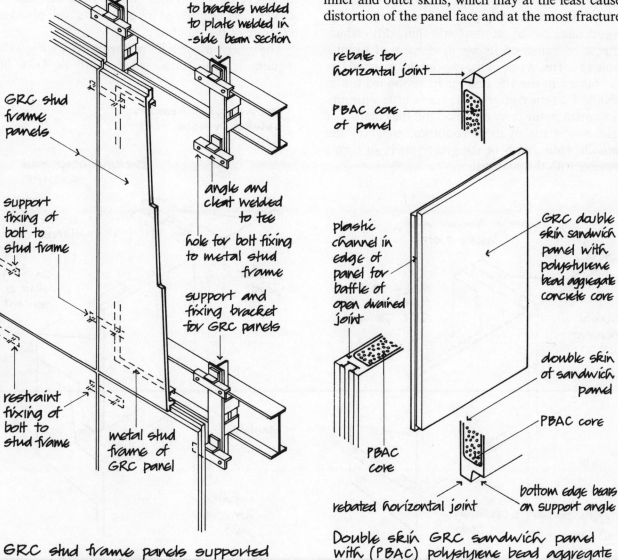

GRC stud frame panels supported and fixed to structural steel frame

Fig. 154

Double skin GRC sandwich panel with (PBAC) polystyrene bead aggregate concrete core

Fig. 155

of the junction of the skins, interstitial condensation in the core that may cause delamination and distortion of panel faces, the sometimes obvious surface rippling of the outer skin over the core ribs and the inevitable thermal bridge of the solid edge GRC material.

The advantages of GRC as a surface material are that it can be used as a thin, lightweight skin with adequate strength and durability for use as a wall facing material for storey height panels and as undersill or spandrel panels to continuous horizontal windows. Both the tensile and impact strength of GRC diminish with time and allowance is made for this loss of strength in design calculations.

The principal disadvantage of GRC as a material is the considerable moisture movement with changing conditions of atmospheric humidity which impose restraints on its use in climates of varying humidity. This is one of the reasons the material has lost favour in the UK and for its continued use in Middle Eastern countries. There is little moisture movement in the drier climates, and the contrast of light and shade in those countries, enhances the smooth white finish of the great variety of forms possible with this material.

Jointing

The three types of joint used are sealant filled, compression gasket and open drained joint. As with other wall panel materials the sealant filled joint illustrated in Fig. 156 is mostly used for smaller panels with joints of uniform width. The face filled joint will require renewal after some years as the mastic oxidises and hardens.

The gasket joint, adopted from glazing techniques, illustrated in Fig. 157, is effective where accuracy in the manufacture and fixing of the panels will ensure a tight fit for the gasket to exclude rain. Preformed cross-over gaskets are heat welded to straight lengths of gasket on site. These compression or push fit gaskets may be recessed for protection or exposed as a feature of the wall face.

The open drained joint illustrated in Fig. 158 requires a considerable edge depth of GRC to

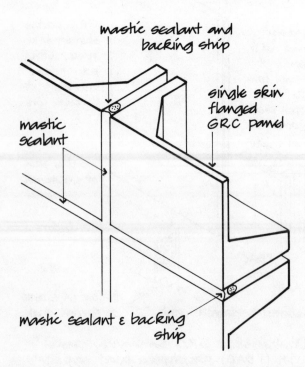

Mastic sealant joint between GRC panels

Fig. 156

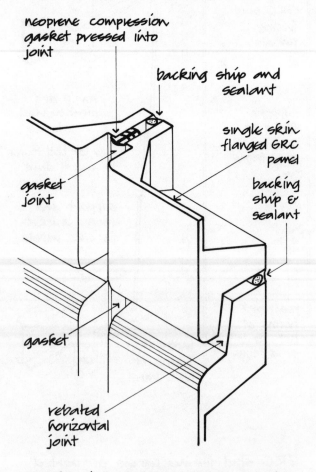

Gasket joint between GRC panels

Fig. 157

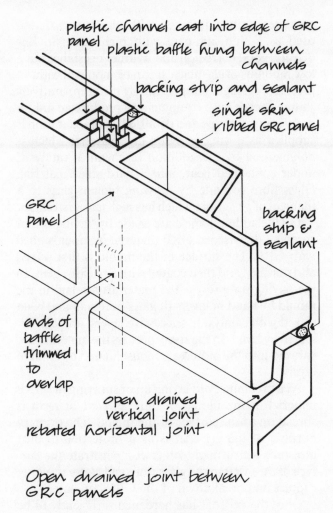

plastic channel cast into edge of GRC panel

plastic baffle hung between channels

backing strip and sealant

single skin ribbed GRC panel

GRC panel

backing strip & sealant

ends of baffle trimmed to overlap

open drained vertical joint

rebated horizontal joint

Open drained joint between GRC panels

Fig. 158

friction pads or, where two support fixings are used, one should allow movement. There should be as few fixings to the carrier system or structural frame as necessary to support and restrain the panel in position and resist lateral wind pressure. For single skin flanged, ribbed and stud frame panels and sandwich panels there should preferably be two lower edge support and restraint fixings and two top restraint fixings to accommodate vertical, horizontal and rotational movements of the panel relative to the frame or carrier system.

The weight of panels is supported on galvanised steel, stainless steel or non-ferrous angles under a GRC flange in the lower edge of panels, as illustrated in Fig. 159, on metal levelling shims or low friction pads or both. The angle is fixed with expanding bolts back to the structure or bolted to carrier systems. Restraint is provided by a dowel, welded to the angle, that fits into a hole or slot in the panel inside a resilient bush or sleeve that allows for movement of the panel.

Restraint fixings are made by threading studs to accommodate the channels which are moulded into edges of panels for the baffle. This deep, thick edge which can be moulded as either a flange, a thickened rib or an edge to sandwich panels acts as a comparatively wide thermal bridge. Open drained vertical joints are used with rebated horizontal joints.

None of these joints will be effective unless there is adequate dimensional accuracy in GRC panels and accuracy in fixing to ensure reasonable uniformity of joint width and a smooth finish to the edges of panels.

Support and restraint fixing of panels

GRC panels should be supported at or near the base of the panel so that the material is in compression to minimise visible shrinkage and movement cracks on the face. The support fixing should either allow some horizontal movement of the panel by the use of low

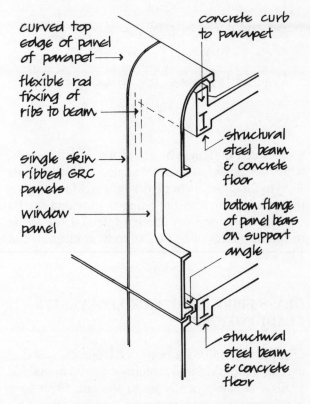

curved top edge of panel of parapet

concrete curb to parapet

flexible rod fixing of ribs to beam

structural steel beam & concrete floor

single skin ribbed GRC panels

window panel

bottom flange of panel bears on support angle

structural steel beam & concrete floor

GRC panels fixed to structural steel frame

Fig. 159

151

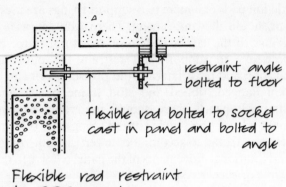

Flexible rod restraint to GRC panel

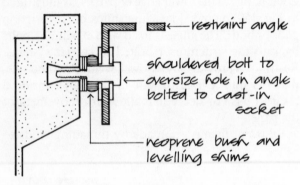

Restraint fixing for GRC panels

Fig. 160

sockets cast in flanges or ribs at the back of panels, which are bolted to restraint angles, bolted to cleats fixed to the structure or carrier system. Movement of the panel relative to the fixing is provided for by oversize holes or slots in the angle and rubber or neoprene bushes or by flexible rods, as illustrated in Fig. 160.

GLASS FIBRE REINFORCED POLYESTER CLADDING (GRP)

Glass fibre reinforced polyester laminate was first used as a thin skin wall cladding material in the mid 1950s and subsequently up to the late 1970s as a lightweight skin panel material for wall systems. This material has never been used as extensively as precast concrete and of recent years its use has declined.

GRP is a composite of a durable, thermosetting polyester resin, reinforced with glass fibre, that is used as a thin laminate with high strength, low density, good corrosion and weather resistance but a low modulus of elasticity. It can be moulded without pressure or high temperature with comparatively simple, inexpensive equipment to produce an unlimited variety of shape and detail. The polyester resin is supplied as a viscous syrup-like material which is polymerised by the addition of chemical catalysts, under controlled heat, into a hard solid material. Glass fibre is made by drawing molten glass to a filament of glass fibre which has high tensile strength.

GRP cladding panels are made by 'laying up' or spraying the viscous GRP material in moulds lined with GRP. The surface of the mould is first waxed and polished and then coated with a release agent. In the 'laying up' process the materials are laid in the mould by hand in layers of glass fibre mat and resin mixed with catalyst in successive layers, and consolidated by hand. In the spray process the materials are sprayed into the mould and consolidated by hand in layers.

As a preliminary to laying up or spraying the GRP material in the mould, a thin gel coat of resin is spread on to the surface of the mould. The primary purpose of the gel coat is as a protection against moisture which might otherwise penetrate the surface of the GRP to the glass fibre and cause swelling, rupture and breakdown of the GRP laminate.

Once the gel coat has hardened sufficiently to be tacky, the first layer of resin, catalyst and glass mat is spread in the mould and consolidated by roller, followed by successive layers up to the required thickness of the laminate. The moulded panel is then taken from the mould and cured in a box or chamber under controlled conditions of temperature and humidity to develop structural and dimensional stability.

Control of the process of manufacturing GRP panels has a most significant effect on the finished product in use as a wall panel material. Selection of the resin, catalyst, fillers and pigment for a particular purpose, the careful mixing of the materials, skill in application of the materials to form a sound laminate, control of curing and control of the conditions of temperature and humidity in the workshop all have a significant effect on the dimensional accuracy, stability, strength and durability of the finished product.

One of the main reasons for the comparatively limited use of this material is the difficulty of control in manufacture and the failures that have been caused by faulty control and poor workmanship.

The natural colour of GRP is not generally accepted as an attractive finish for wall panels and it is usual to make panels that are coloured with a pigment added to the gel coat or the resin binder. The addition of pigment and the selection of colour can appreciably affect the weathering characteristics of GRP panels. Strong colours such as oranges and reds tend to fade through the effect of ultraviolet light, which causes the surface to chalk, and colour fastness may well be irregular between panels. Dark colours encourage high surface temperatures that increase the risk of separation of the laminae of the skin, a failure that is known as lamination. Variations in the surface of flat smooth faced panels, caused by mechanical and thermal distortion, may be obvious in surfaces more than about a metre wide. A matt texture or shallow ribbed finish will effectively mask distortions of the surface without noticeably affecting the smooth sleek surface of this finish.

To enhance the poor resistance to damage and spread of flame characteristics of GRP it is practice to add fillers or chemical additives to the resin or to coat the surface. The addition of fillers to improve the fire retardancy of the material has the effect of weakening its capacity to resist weathering agents and affects pigments which are added. The addition of fillers and pigment to improve fire resistance and colour appreciably reduces the weathering characteristics of this material. One method that has been used to provide protection against weathering and to improve fire resistance is to coat the surface with polyurethane to improve weathering and to modify the gel coat to improve fire resistance.

Because GRP is an expensive material, it is used as a thin skin for all panels in thicknesses of from 3 to 6, and has to be stiffened with edge flanges, shaped profiles, stiffening ribs or a sandwich construction. To be effective as stiffening, shaped profiles must be deep in relation to the thickness of the skin. Because GRP is used for the dramatic effect of the level face of panels, the usual method of stiffening is to bond stiffening ribs to the back of the panel skin. These stiffening ribs are usually made of hollow sections of GRP that are overlaid with GRP as the laminate is built up. Figure 161 is an illustration of a ribbed panel stiffened with top hat and box section hollow ribs.

In common with the other thin skin panel material, metal, GRP can be formed with ease around a

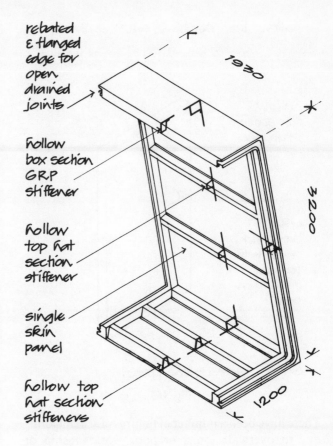

Ribbed GRP panel

Fig. 161

core of insulating material as a sandwich panel. The sandwich of GRP at once provides stiffening and insulation. These sandwich panels are made with two laminate skins of GRP moulded around the insulating core with the GRP skins joined around the edges of the panel to seal the sandwich and for fixing. Figure 162 is an illustration of a sandwich panel. The size of these panels should be limited to avoid too great a distortion of the finished panel face through differential expansion of core and skin material.

Jointing

As with any other facing panel wall structure the joints between the panels of GRP have to allow for differential structural, thermal and moisture movements between the supporting frame and the wall and must serve as an effective barrier to penetration of rain. The three types of joint that have been used are mastic sealant, gasket and open drained joint.

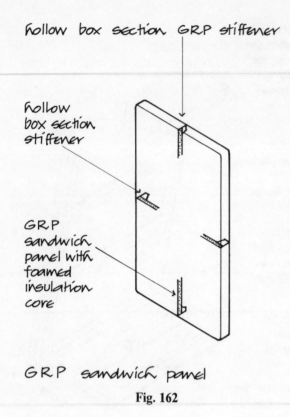

hollow box section GRP stiffener

hollow box section stiffener

GRP sandwich panel with foamed insulation core

GRP sandwich panel

Fig. 162

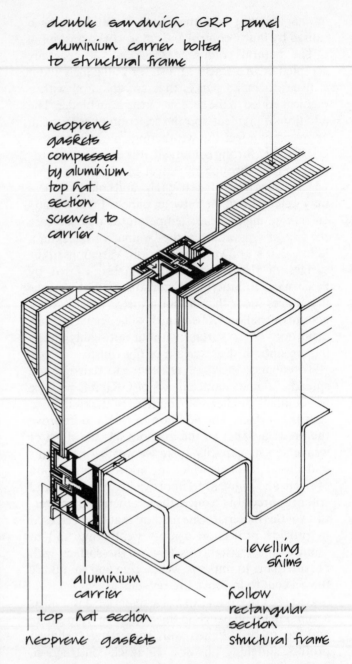

double sandwich GRP panel

aluminium carrier bolted to structural frame

neoprene gaskets compressed by aluminium top hat section screwed to carrier

aluminium carrier

top hat section

neoprene gaskets

levelling shims

hollow rectangular section structural frame

Double sandwich GRP panels in neoprene gaskets fixed to aluminium carrier system

Fig. 163

There have been a number of failures of sealant joints due to overwide joints or poor workmanship or both, which resulted in unsightly mastic failure, and which have given this method of jointing a bad name. Providing the joint is reasonably tight and adequate to the anticipated movements, a skilfully applied sealant joint will give satisfactory performance for the life of the sealant material, which will need periodic attention.

Gasket jointing techniques, adopted from glazing, have been successfully used for GRP panels. Preformed gaskets of neoprene or ethylene propylene diamine monomer (EPDM) fit around and seal the edges of adjacent panels with the gasket compressed to the panels with adjustable clamps bolted to the carrier system, as illustrated in Fig. 163. These gaskets have preformed cross-over intersections at the junction of horizontal and vertical joints that are heat welded on site to straight lengths of gasket. Both neoprene and EPDM gaskets oxidise, harden and lose resilience on exposure and may need replacement every 10 to 20 years.

Open drained joints have the advantage that the jointing material is not visible and that the open drain serves as a check to driving rain. The open drained joint illustrated in Fig. 164 has outer and inner drain channels and a baffle.

Support and fixing

Because GRP is a comparatively expensive material it is used as a thin skin and in consequence GRP panels are lightweight and do not by themselves require substantial support fixings. Because of the

154

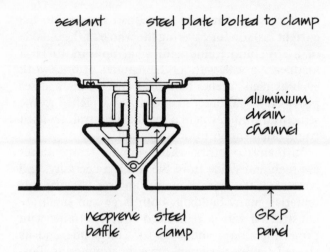

sealant — steel plate bolted to clamp

aluminium drain channel

neoprene baffle — steel clamp — GRP panel

Open drained vertical joint between GRP panels

Fig. 164

considerable thermal movement of GRP and the thin skin form of its use, it is essential to support and restrain panels to or in fixings that will allow for thermal movement and restrain the lightweight panels in position. Usual practice is to clamp panels back to the structure or to the carrier system inside neoprene or EPDM gaskets that hold the panels in place and act as weather seal. The neoprene gasket illustrated in Fig. 163 is clamped in place by a top hat metal section screwed to the aluminium carrier.

Another method of support and fixing is to incorporate timber battens or framing in either single skin or sandwich panels. The timber battens can be enclosed in the GRP laminate as stiffening ribs and used as a means of fixing the panel by screwing back to the carrier system or structure and as a means of fixing for windows.

GRP as a material for wall panels lost favour principally because of failure due to poor manufacturing techniques and lack of colour fastness to expose surfaces. Since the introduction of this material for use as wall panels there have been considerable improvements in the mixing manufacture and use of thermosetting materials which, applied to this type of panel, could make it wholly acceptable as a wall panel material.

INFILL WALL FRAMING TO A STRUCTURAL GRID

Infill wall frames are fixed within the enclosing members of the structural frame or between projec-

tions of the frame, such as floors and roof slabs, which are exposed as illustrated in Fig. 165. The infill wall may be framed with timber or metal sections with panels of glass in the form of a window wall, framed around solid panels of any one of the thin sheet materials, framed above a brick or block riser

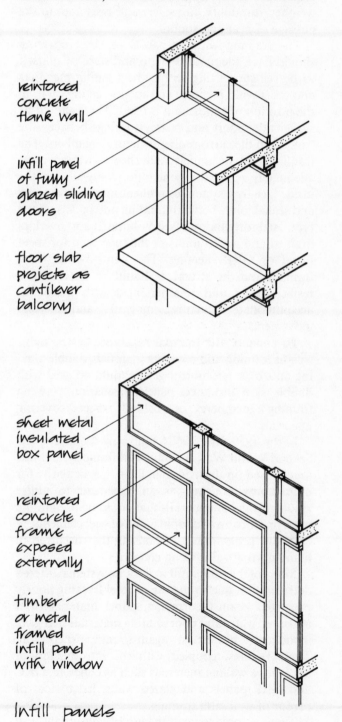

reinforced concrete flank wall

infill panel of fully glazed sliding doors

floor slab projects as cantilever balcony

sheet metal insulated box panel

reinforced concrete frame exposed externally

timber or metal framed infill panel with window

Infill panels

Fig. 165

below sill level or framed as support for wood, metal or GRP sheeting with or without windows. Any one or more of the thin sheet materials may be used as panel framing or covering to the supporting frame to serve as a wall element to satisfy the functional requirements of strength and stability, resistance to weather, durability, fire safety and resistance to the passage of heat and sound.

The framing with its panels or sheet covering should have adequate strength and stability in itself to be self-supporting within the framing members and resist wind pressure and suction acting on it, in the position of exposure it is fixed. There should be sufficient support and restraint fixings between the frame and the surrounding structural members. The framing and its panels and sheet covering must adequately resist the penetration of water to the inside face by a system of resilient mastic or drained and sealed joints between framing and panels, of the type used for windows (see Volume 2) and overlap, drained and sealed joints of the type used for sheet metal or GRP sheeting. The joints between the framing and the structure should be filled with a resilient filler and weathersealed with mastic to accommodate structural, moisture and thermal movements.

To enhance the thermal resistance of the lightweight framing and covering materials double glazing and/or solar control glass should be used with double skin insulated panels, insulation between framing members or behind sheet covering materials.

In the 1950s and 1960s, following the end of the Second World War, the infill wall frame system was much used in framed buildings, particularly for multi-storey housing, as an expedient to utilise readily available materials that could be used in the mass production and rapid fixing of wall elements, in extensive programmes of housing that occurred in many northern European countries.

Many of the early infill wall frame systems suffered deterioration due to the use of steel framing poorly protected against corrosion, panel materials that absorbed water and poor jointing materials that gave inadequate protection against rain penetration. These failures, coupled with the introduction of alternative walling materials such as concrete, GRC and GRP panels and glazed walls, led to loss of favour of wall infill framing.

There is some logic in the use of lightweight wall framing as an apparent cladding element within the load carrying structural frame, as compared to the current fashion for covering the whole of the outside of a structural frame with what appears to be a loadbearing wall with all the features of brickwork such as piers, arches and decorative string courses provided by an outer skin of brick attached to the structural frame, where a loadbearing wall by itself could often support floors by itself.

In many countries where summer temperatures are high and shade from the sun is a necessity, and concrete is the most economic and readily available material, many buildings both large and small are constructed with a reinforced concrete frame with projecting floors and roof for sun shade and as shaded outdoor balcony areas in summer as illustrated in Fig. 165. Because of the protection afforded by the projecting floor slabs and roof against wind driven rain and the diminuition of daylight penetration caused by these projections, in winter months, it is common practice to form fully glazed infill panels in this form of construction.

With the variety of solid walling materials, such as brick, stone and block, and lightweight panels and wall sheeting available extensive combinations of these materials are possible as infill walling to framed structures.

Infill wall framing, which is currently out of fashion may well, as fashions do, have a new lease of life as a sensible form of enclosure to buildings.

GLAZED WALL SYSTEMS

Up to the beginning of the twentieth century glass was a comparatively expensive material. Window glass was made by hand in the spun, crown glass process and later the blown cylinder process. Window glass made by these processes was cut into comparatively small squares (panes) for use in the windows of traditional loadbearing walls. Plate glass was made by casting, rolling and grinding and polishing sheets of glass both sides. These laborious methods of production severely limited the use of glass in buildings.

With the development of a continuous process of drawing window glass in 1914 and a process of continuously rolling, grinding and polishing plate glass in the 1920s and 1930s, there was a plentiful supply of cheap window glass and rolled and polished plate glass.

In the 1920s and 1930s window glass was extensively used in large areas of windows framed in slender steel sections as continuous horizontal features between under sill panels and as large metal framed windows. During the same period rolled plate glass was extensively used in rooflights to factories, the glass being supported by glazing bars fixed down the slope of roofs. Many of the sections of glazing bar that were developed for use in rooflights were covered by patents so that roof glazing came to be known as 'patent glazing' or 'patent roof glazing'.

The early uses of glass as a wall facing and cladding material were developed from metal window glazing techniques or by the adaptation of patent roof glazing to vertical surfaces, so that the origins of what came to be known as 'curtain walling' were metal windows and patent roof glazing.

The early window wall systems, based on steel window construction, lost favour principally because of the rapid and progressive rusting of the unprotected steel sections that in a few years made this system unserviceable. The considerable buckling and distortion of frames and fracture of glass that was due to rusting, rigid putty fixing of glass and rigid fixing of framing gave steel window wall systems a bad name.

With the introduction of zinc coated steel window sections and the use of aluminium window sections there was renewed interest in metal window glazing techniques. Cold formed and pressed metal box section subframes, which were used to provide a bold frame to the slender section of metal windows, were adapted for use as mullions to glazed wall systems based on metal window glazing techniques. These hollow box sections were used either as mullions for mastic and bead glazing of glass and metal windows or as clip on or screw on cover sections to the metal glazing.

Hollow box section mullions were either formed in one section as a continuous vertical member, to which metal window sections and glass were fixed, or as split section mullions and transomes in the form of metal windows with hollow metal subframes that were connected on site to form split mullions and transomes. The complication of joining the many sections necessary for this form of window panel wall system and the attendant difficulties of making weathertight seals to the many joints has, by and large, led to the abandonment of window wall glazing systems.

Glass for rooflights fixed in the slope of roofs is to a large extent held in place by its weight on the glazing bars and secured with end stops and clips, beads or cappings against wind uplift. The bearing of glass on the glazing bars and the overlap of bays down the slope act as an adequate weather seal.

To adapt patent roof glazing systems to vertical glazed walls it was necessary to provide a positive seal to the glass to keep it in place and against wind suction, to support the weight of the glass by means of end stops or horizontal transomes and sills and to make a weathertight seal at horizontal joints.

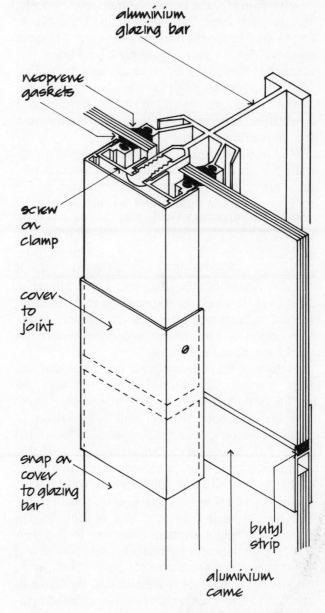

Aluminium glazing bar used for vertical glazing

Fig. 166

The traditional metal roof glazing bar generally took the form of án inverted 'T' section with the tail of the T vertical for strength in carrying loads between points of support with the two wings of the T supporting glass. For use in vertical wall glazing it was often practice to fix the glazing bars with the tail of the T inside with a compression seal on the outside holding the glass in place, as illustrated in Fig. 166.

The usual section of metal glazing bar, which is well suited to roof glazing, does not provide a simple, positive fixing for the horizontal transomes and sills necessary for vertical glazing systems. The solution was to use continuous horizontal flashings or cames on to which the upper bays of glass bore and up to which the lower bays were fitted, as illustrated in Fig. 166. Patent roof glazing techniques, adapted for use as vertical glazing, are still in use but have by and large been superseded by extruded hollow box section mullion systems.

Hollow box section mullions were designed specifically for glass curtain walling. These mullion sections provided the strong vertical emphasis to the framing of curtain walling that was in vogue in the 1950s and 1960s and the hollow or open section transomes with a ready means of jointing and support for glass. The pattern of what came to be known as curtain walling was set by the United Nations Secretariat Building and Lever House in New York, in which the framing elements of slender vertical mullions supporting glass and smooth panels were emphasised by mullions as continuous verticals up the height of the building.

Hollow box section mullions, transomes and sills were generally of extruded aluminium with the section of the mullion exposed for appearance sake and the transome and sill and head joined to mullions with spigot and socket joints, as illustrated in Fig. 167. A range of mullion sections was available to cater for various spans between supporting floors and various wind loads. The mullions, usually fixed at about 1 to $1\frac{1}{2}$ metre centres, were secured to the structure at each floor level and mullion lengths joined with spigot joints as illustrated in Fig. 167. The spigot joints between mullions and mullions and between mullions and transomes, head and sill, made allowance for thermal movement and the fixing of mullion to frame allowance for differential structural, thermal and moisture movements. Screw on or clip on beads with mastic or gasket sealants held the glass in place and acted as a weather seal. This form of curtain walling with exposed mullions was the

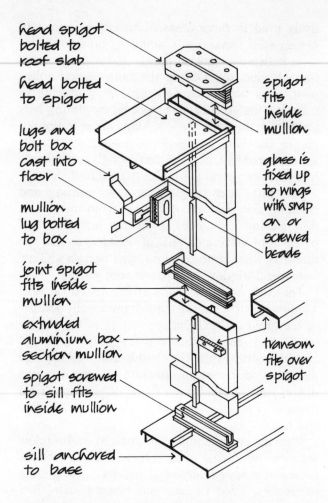

head spigot bolted to roof slab

head bolted to spigot

lugs and bolt box cast into floor

mullion lug bolted to box

joint spigot fits inside mullion

extruded aluminium box section mullion

spigot screwed to sill fits inside mullion

sill anchored to base

spigot fits inside mullion

glass is fixed up to wings with snap on or screwed beads

transom fits over spigot

Extruded aluminium curtain walling

Fig. 167

fashion during the 1950s, 1960s and early 1970s.

Since then fashion has changed. The introduction of tinted solar control glass and the use of gaskets to provide a more positive rain and wind weather seal around glass has facilitated a move to systems of glass walls where the hollow box section framing members are fixed behind the glass, which is held in slender gaskets, to give the appearance of a glass wall, as illustrated in Figs 168 and 169.

More recently the use of toughened glass, hung from brackets fixed to the structure, has provided the means of effecting what is truly a curtain wall of glass. The large squares of toughened glass are hung from the top by metal studs anchored to the frame with additional restraint fixings, as illustrated in Fig. 170. The joints between the glass are sealed with a silicone based sealant.

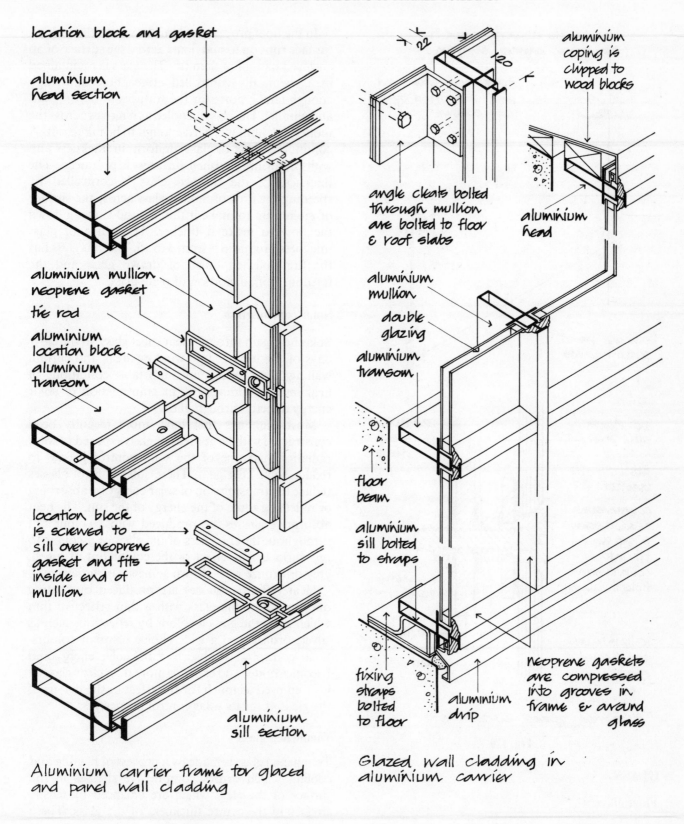

location block and gasket

aluminium head section

aluminium mullion neoprene gasket

tie rod

aluminium location block aluminium transom

location block is screwed to sill over neoprene gasket and fits inside end of mullion

aluminium sill section

Aluminium carrier frame for glazed and panel wall cladding

22

120

aluminium coping is clipped to wood blocks

angle cleats bolted through mullion are bolted to floor & roof slabs

aluminium head

aluminium mullion

double glazing

aluminium transom

floor beam

aluminium sill bolted to straps

fixing straps bolted to floor

aluminium drip

neoprene gaskets are compressed into grooves in frame & around glass

Glazed wall cladding in aluminium carrier

Fig. 168

Fig. 169

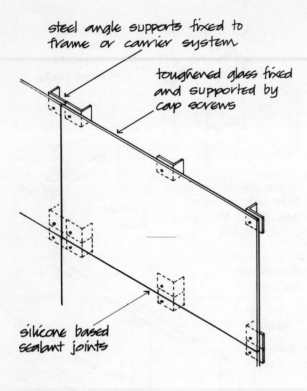

steel angle supports fixed to frame or carrier system

toughened glass fixed and supported by cap screws

silicone based sealant joints

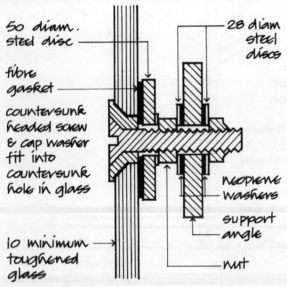

50 diam. steel disc

fibre gasket

countersunk headed screw & cap washer fit into countersunk hole in glass

10 minimum toughened glass

28 diam steel discs

neoprene washers

support angle

nut

Toughened glass curtain wall

Fig. 170

GLASS

Float glass

Most of the flat glass used in building today in the UK is produced by the float method of production that was first introduced in 1959.

In the float process, molten glass from the melting furnace runs on to and floats across the surface of an enclosed bath of molten tin. The glass is maintained in a chemically controlled atmosphere at a high enough temperature for the surfaces to become flat and parallel. The glass is cooled as it moves across the molten tin, until it is hard enough to be taken out. A continuous sheet of glass, uniform in thickness and with bright fire polished surfaces, is produced. The thickness of the finished glass is controlled by speeding the flow of molten glass across the surface of molten tin to make it thinner and slowing down the flow to make it thicker. The range of glass thickness produced is from 3 to 15. This flat glass has the fire polished finish of drawn glass and the freedom from distortion of plate glass.

Solar control glass

Solar heat gain through clear sheet glass, in the early days of the use of large window areas and curtain walling, did produce uncomfortable conditions of heat inside buildings by the transmission of solar energy directly through glass.

Most buildings that have more recently been constructed with large areas of glass exposed to solar radiation, use one of the solar control glasses to reduce solar heat gain. These solar control glasses reduce the transmission of solar energy by absorbing or reflecting some of the energy of the sun. The heat absorptive glasses are produced with a colour tint throughout the thickness of the glass or a colour to one surface. The effect of the colour tint of green, grey or bronze is to absorb some solar radiation.

Heat reflective glasses are produced by coating one surface of the glass with a thin reflective film which gives the glass a colour by reflection, such as silver, bronze and a wide choice of other colours. Solar control glass absorbs more solar energy and becomes hotter and expands more than clear glass. It is often used as much for the effect of the colour of the glass as for its solar control property.

Toughened glass

Toughened glass is made by a process of heating and cooling which causes compressive stresses in the surface of the glass which are balanced by tensile stresses in the centre thickness of the glass. These counterbalancing stresses give toughened glass its increased strength. This glass will only break under extreme loads which bend the glass sufficiently to overcome the stresses, or by severe impact with a sharp object that may penetrate the surface and so release the stresses and cause the glass to fracture. Toughened glass is up to five times stronger than ordinary glass of the same thickness.

The proprietary names 'Armourplate', 'Armourcast' and 'Armourclad' are used for toughened glass.

A glazed wall system should satisfy the functional requirements of a wall where it serves as the building envelope, and the requirements of a window where it serves to admit daylight and provide ventilation.

Strength and stability

The strength and stability of a glazed wall depend on the size, thickness and nature of the glass in resisting lateral wind pressure, the section of the mullions between support and restraint fixings to the structure, the section of the transomes and sills in supporting glass and the arrangement of joints between framing members and between glass and framing member, to allow for thermal movement of the glass wall and structural movements.

For economy in the use of glass and framing, the usual spacing of proprietary mullion sections is from 760 to 1200 for 6 thick float glass. Mullion spacing of up to 2400 has been used with mullions specifically engineered for the purpose. Mullions should have adequate section to withstand lateral wind pressures acting on the glass they support between fixings at each floor level without undue deflection and they should have sufficient strength to support the weight of glass and framing between support fixings without undue lengthening by elastic strain. The system of support fixings for mullions depends on the arrangement of panels of glass, both fixed and for opening lights and opaque panels of glass or sheet metal between floors. Usual practice is to use one support fixing with one restraint fixing for each storey height or one support fixing and two restraint fixings for every two storey height lengths of wall so that the mullions are hung from a rigid fixing and restrained by lower fixings to allow for differential movements between structure and glass wall framing.

Support fixings take the form of lugs or brackets fixed to the back of mullions that are bolted to the structure, as illustrated in Fig. 169, with expanding bolts fixed to holes drilled in concrete or directly to steel. Each support fixing must be firmly attached to the structure to support the necessary weight of glass

and framing. No provision for movement is made at support fixings.

Restraint fixings are designed to retain the framing and allow for differential movements between the glass wall system and the structure. Restraint fixings take the form of lugs or brackets fixed to the back of mullions and bolted to the structure, through slotted holes to allow some differential movement, with low friction plastic washers that allow movement and at the same time restrain the fixing. Where there are support fixings at each floor level for storey height lengths of mullion, the spigot joint between mullions illustrated in Fig. 167 serves as a restraint fixing. Another form of restraint fixing is a bracket fixed to the structure where the back of the mullion is restrained in a channel in which the mullion has some freedom of vertical movement.

Movement joints between frame members are formed by spigot joints between lengths of mullion and spigot joints of transomes and sills to mullions, such as those illustrated in Fig. 167. These joints, which are weather-sealed with mastic or with neoprene gaskets (Fig. 167), allow for thermal movements of glass and framing and some differential movement between glazed wall and structure.

Glass is secured in place on spacer blocks under the lower edge of the glass and with neoprene gaskets that are compressed to the glass by a compression fit, as illustrated in Fig. 169, or by a compression strip on the gasket, as illustrated in Fig. 166. The edge clearance of the glass in the framing and the resilience of the gasket allow for thermal movement of the glass in the metal framing.

Exclusion of wind and rain

The smooth, hard impermeable surface of glass in curtain walls allows wind-driven rain to flow in sheets both across and up and down the face of the wall. Rain, under pressure of wind, will penetrate the smallest gap at joints between glass and framing. The joints between glass and framing have to be wide enough to accommodate movements and at the same time serve as a weather seal to both wind and rain.

Many of the earlier curtain wall systems relied on cover beads to keep the glass in place and mastic seals to exclude rain, with a drainage channel behind the glass to collect rain that penetrated joints. The most vulnerable points in these systems were the junctions of horizontal to vertical framing members, where the joints between beads and frame members generally

161

allowed some penetration of rain and wind. Current practice is to employ neoprene or EPDM gaskets preformed to fit around each individual square of glass that is fixed close to the outside face of the mullion. These gaskets effectively seal the junction of glass and framing. The neoprene gasket illustrated in Fig. 169 is a push fit compression seal that is compressed into a rebate in the frame and fits tightly around the edges of the glass. The seal illustrated in Fig. 166 is a compression seal that fits around the glass and is compressed into a frame with a screwed on compression strip. Neoprene gaskets have better resilience but become hard and brittle more quickly than EPDM seals. On exposure to sunlight, both types of seal oxidise, harden and lose resilience and may need replacing every twenty years or so. Compression gaskets preformed as a continuous joint around each square of glass are the most effective weatherseal.

Thermal properties

A thin sheet of glass provides negligible resistance to the transfer of heat. Systems of double or triple glazing can be used to improve thermal properties of glass. The transfer of heat of single glazing is assumed as 5.7 Wm^2K, double glazing as 2.8 W/m^2K and triple glazing as 2.0 W/m^2K. The improvement in insulation by using double glazing is plainly worthwhile. Building regulations concerned with conservation of energy require a minimum level of insulation against transfer of heat through walls and permit a percentage of the area of the wall to be glazed. The requirements for offices is that a maximum of 35% of the area of a wall may be single glazed. A proportionally larger area may be double glazed. The effect of this regulation is that the rest of the wall requires an insulation value of 0.45 W/m^2K. The consequence is that either nearly two-thirds of a glass wall have to have some form of insulating lining or backing to comply with the regulations, or some system of heat recovery has to be used to satisfy the energy conservation requirement. As an alternative the calculation procedure may be used to certify that the energy consumption in a fully glazed building would be no more than it would be for a similar building, calculated by the elemental approach.

For the purposes of insulation, most glass wall systems use double glazing with composite metal panels between sill and head levels of windows or inner linings of insulting panels or back up walls or combinations of these. Many glazed wall systems use tinted solar control glass either as single or double glazing.

Thermal bridge

The metal framing members of glazed wall systems are good conductors of heat and act as a thermal bridge. Where double glazing and insulated panels are used the thermal bridge effect of the supporting metal frame is such that it is worth considering the use of some form of thermal break. The simplest form of thermal break is made by extending the neoprene gasket system over the face of the metal frame. This provides some little reduction in transmittance. A more sophisticated and more effective thermal break is to interpose some material with low transmittance in the framing as a break in the thermal bridge. The plastic thermal break in the carrier system illustrated later, in Fig. 177, is effective as a break in the thermal bridge of the frame, which itself has an insulated core. Plainly the complication of this form of construction is only worthwhile in framing members around panels with moderate or good insulating values.

Acoustic properties

A thin sheet of glass offers little resistance to the transmission of airborne sound because of its small mass per unit area. Double or triple glazing will effect some reduction in sound transmission (see Volume 2).

Fire resistance

A glass wall system is an 'unprotected area' as defined in the Building Regulations and as such is limited in area for buildings over 15 metres high. In multi-storey glazed wall systems it is necessary to use some form of fire resistant panel, lining or back up wall to a part of glazed walls. This requirement for fire resistance is usually combined with the requirement for thermal insulation.

SHEET METAL WALL CLADDING

Sheet metal has for many years been used as wall cladding principally in the form of profiled sheets of steel or aluminium that are used both as a roof covering and wall cladding to factories and other single storey buildings. The original sinusoidal section corrugated iron has largely been replaced by trapezoidal section sheets, for their improved span and appearance. Both galvanised steel and aluminium profiled sheets can be coated with an inorganic plastic coating as a protection and for the range of colours possible with these coatings. The properties, sections and uses of these profiled sheets were described in Volume 3.

Because of the very poor thermal properties of a thin sheet of metal and current requirements for energy conservation, a range of composite metal sheets is produced in which an insulating core is sandwiched between two sheets of metal that are profiled to provide stiffness. These composite sheets are used for roof and wall cladding to single storey buildings, for both the advantage of the insulation core and the internal lining of sheet metal.

Sheet metal in the form of panels has been in limited use for some 50 years, principally in France, Germany and America, both as wall cladding and more commonly as opaque panels to curtain wall framing. Of recent years sheet metal cladding panels have been more extensively used, in the vogue for 'high technology' applied to buildings, for the smooth, flat and curved surfaces associated with the modern rolled or pressed metal sheet, machine age image.

Sheet metal cladding can be broadly grouped as:

- Single skin and composite sheets
- Flat and profiled single skin panels
- Flat and profiled composite panels

Single skin profiled sheets

Single skin profiled sheets are made from a steel or aluminium strip that is cold rolled in one direction to a standard range of sinusoidal and trapezoidal profiles. The strength of these sheets depends on the depth of the one-way profile in supporting the loads common to roofs and walls. Because the strength of these sheets lies in the direction of the profile, they are supported by and fixed to sheeting rails at right angles to the profile. As roof and wall covering, the sheets are fixed with end laps of sheets and side laps of the profile to exclude rain. For insulation, an underlining of insulating material or a sandwich of insulating material with an underlining sheet is used. The use of these sheets as roof and wall cladding was described in Volume 3.

Profiled sheeting has traditionally been used as roof and wall cladding to single storey buildings in what is sometimes referred to as the 'shed form' of building common to factories, warehouses and other such buildings with large areas of roof and walls, often without windows. A limited range of flashings is available for weathering and to provide a finish at sills, eaves, ridges and around openings.

Of recent years profiled sheets have been used in non-traditional ways in what has come to be known as the 'super shed' form, principally for single storey buildings. Profiled sheets have been used with the profile fixed horizontally or diagonally across walls as a continuous wall covering, often with curved sheets to eaves and corners, or in panels with purpose made angles both of metal and other sheet material, with the structural frame either inside the enclosing cladding or exposed outside for effect.

Composite sheets

A range of standard, flat and profiled composite sheets is available with an insulating core between two sheets of steel or aluminium to combine an acceptable internal finish with adequate insulation. The advantage of profiled composite sheets is in the economy of continuous rolling for mass production. These sheets which are produced in lengths of up to 10 metres, are designed for face fixing to sheeting rails fixed to the structure. The difficulty of making a neat, weathertight and attractive finish to the end of the one way profile of these sheets at corners and around openings has limited their use to simple, wide span shed forms of buildings. The profiled, composite steel sheeting, illustrated in Fig. 171, is designed for use as side wall sheeting to single storey buildings where the sheet can be used in single lengths without horizontal joints and with a side lap formed by the profile.

Flat composite sheets with secret fixings are designed to combine the economic advantages of continuous rolling and the stiffness provided by lamination of the insulating core to the two skins, with the appearance of flat panels for use as walling. The flat steel sheets illustrated in Fig. 172 are made in widths of 900, lengths of up to 10 metres and thickness of 50. The long edges of the sheets are formed to provide an interlocking joint in which the fixings to the sheeting rails are hidden. Sheeting rails are required at up to 3 metre intervals. These sheets are galvanised and coated with coloured inorganic

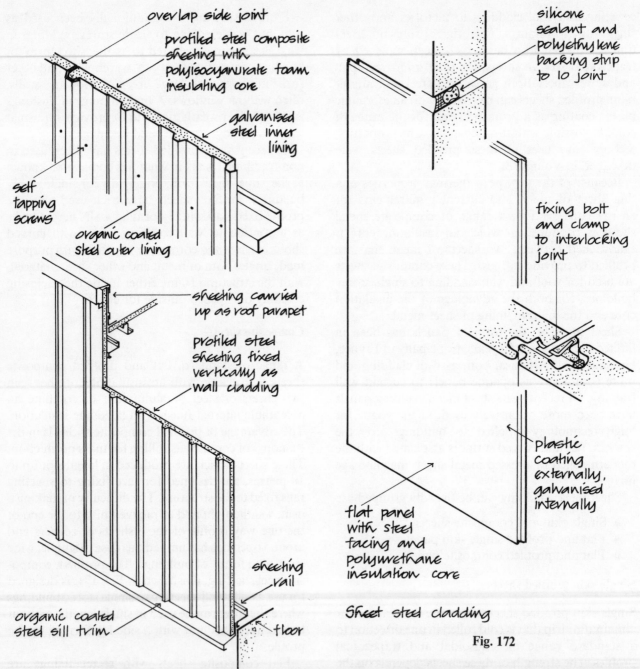

overlap side joint

profiled steel composite sheeting with polyisocyanurate foam insulating core

galvanised steel inner lining

self tapping screws

organic coated steel outer lining

sheeting carried up as roof parapet

profiled steel sheeting fixed vertically as wall cladding

organic coated steel sill trim

sheeting rail

floor

Profiled composite metal sheeting

Fig. 171

silicone sealant and polyethylene backing strip to lo joint

fixing bolt and clamp to interlocking joint

plastic coating externally, galvanised internally

flat panel with steel facing and polyurethane insulation core

Sheet steel cladding

Fig. 172

coating externally and painted internally. End joints are sealed with mastic as illustrated. These sheets can be made with openings for windows with square or rounded corners. The principal use of these flat sheets is for continuous undersill panels between continuous horizontal windows and as flat cladding to walls of single storey buildings.

The aluminium faced composite sheet illustrated in Fig. 173 is made in widths of 600, lengths of up to 10 metres and thickness of 50. Horizontal and vertical joints are made with an insulated plastic insert that is fixed in the edge of one sheet and fits to the adjacent sheet as a male and female joint. The joint is sealed with a neoprene gasket as illustrated. Window panels and square and rounded corner fittings are supplied. The exterior face of these sheets is finished with an oven dried paint finish in a range of colours.

164

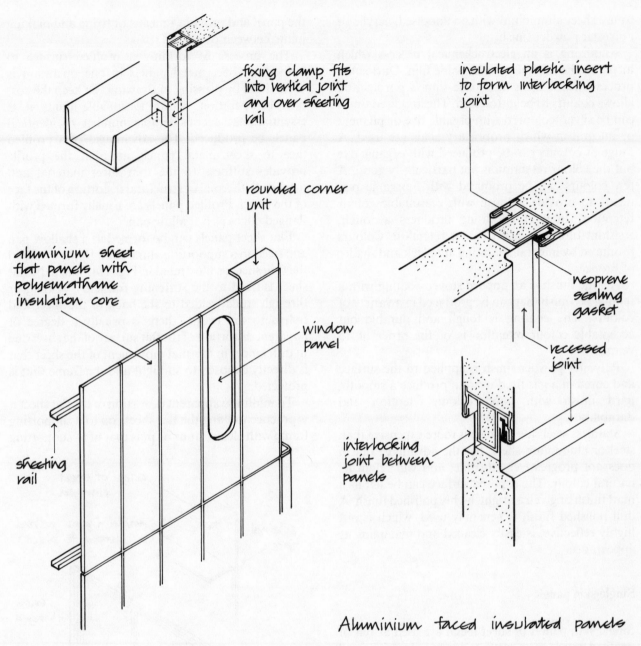

fixing clamp fits into vertical joint and over sheeting rail

rounded corner unit

aluminium sheet flat panels with polyeurathane insulation core

window panel

sheeting rail

insulated plastic insert to form interlocking joint

neoprene sealing gasket

recessed joint

interlocking joint between panels

Aluminium faced insulated panels

Fig. 173

SHEET METAL WALL PANELS

Sheet metal is used in the form of separate flat or profiled panels supported by a metal carrier system which is fixed to the structure or the panels are fixed to rails fixed to the structure. These panels are separated either by visible joints between panels or separated by the visible members of their supporting frame.

The materials used for metal panels are steel sheet,

aluminium sheet and stainless steel sheet. Galvanised steel sheet is usually coated on the exposed face with one of the inorganic plastic coatings described in Volume 3.

Aluminium sheet is the most used material for sheet metal panels because the metal is not liable to progressive corrosion. On exposure to atmosphere a thin, stable oxide forms as a coarse textured coating that inhibits further corrosion. Because the natural oxide film has a dull colour and texture it is common to use sheet aluminium with an anodised, acrylic or polyester powder finish.

Anodising is an electrochemical process which increases the thickness of the oxide film. During the process of anodising the oxide film is porous and allows colours to be introduced. The anodised finish can be silver, coloured with organic dye or pigment or integral in which proprietary acids are used. A range of colours can be produced with organic dye but the colour retention is not particularly good. A few colours can be produced with inorganic pigments to provide a finish with reasonable colour retention. Integral anodising produces a tough, resistant film with good colour retention. Colours produced by integral anodising are black and shades of bronze.

Acrylic finish is an applied stoved coating with a semi-gloss finish that can be produced in a variety of colours. The coating is tough and durable but acceptable colour retention is of the order of 15 years.

Polyester powder finish is applied to the surface and cured in a gas fired oven to produce a smooth, hard finish with good colour retention and durability.

Stainless steel sheet, which is more expensive than steel or aluminium sheet, has the advantage that it does not progressively corrode, and will retain its natural colour. The exposed surface can be left as a matt finish or given a bright, highly polished finish. A dull polished finish is generally used, which is not highly reflective, is easily cleaned and maintains its appearance.

Single skin panels

Single skin panels of sheet metal are used as flat or profiled panels as an outer facing to lightweight wall cladding systems. The thin sheet material is stiffened by forming it as a shallow pan, by pressing or drawing to a profiled finish or by fixing or suspending flat panels on a carrier frame.

The process of forming sheet metal as a shallow pan is a comparatively simple and inexpensive process. The corners of the sheet are cut, the edges cold bent using a brake press and the corners are then welded. A wide variety of edge profiles can be produced by brake pressing. The advantages of forming sheet metal panels as a shallow pan are that the flanged pan edges provide a degree of stiffness to

the panel and provide a means of fixing and making joints between panels.

The process of forming a profiled surface to individual sheet metal panels in one operation is carried out by pressing or drawing. To keep the cost of this operation within reasonable limits it is essential that a considerable number of identical panels be produced. The advantages of a profiled face to sheet metal panels are that the profile provides stiffness to the thin sheet material and masks any thermal or structural distortion of the face of the panel. Profiled panels are usually formed with flanged edges as a shallow pan.

Flat sheet panels can be formed as a shallow pan and fixed to a supporting aluminium frame to stiffen the thin sheet as illustrated in Fig. 174. When the flat sheet is fixed to the stiffening frame by welding or through studs welded to the back of the sheet and bolted to the frame, there is usually a degree of apparent distortion of the flat surface of the sheet due to differences in thermal movement of the sheet that is directly exposed to sunlight and the frame that is protected.

To minimise apparent distortion of the flat sheet it is practice to hang the flat sheets on to a supporting frame with cleats, studs or pins that fit to supporting

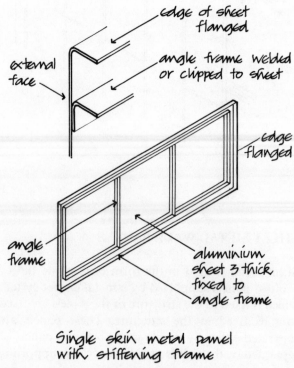

edge of sheet flanged

angle frame welded or clipped to sheet

external face

edge flanged

angle frame

aluminium sheet 3 thick fixed to angle frame

Single skin metal panel with stiffening frame

Fig. 174

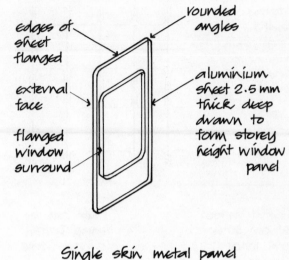

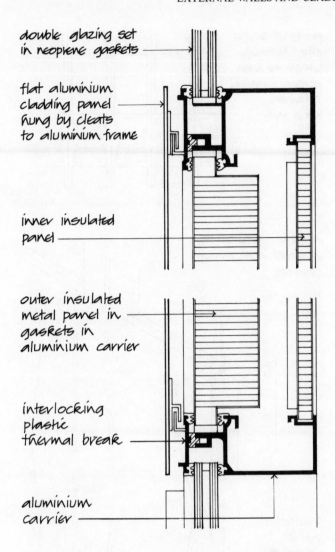

double glazing set in neoprene gaskets

flat aluminium cladding panel hung by cleats to aluminium frame

inner insulated panel

outer insulated metal panel in gaskets in aluminium carrier

interlocking plastic thermal break

aluminium carrier

Flat single skin aluminium panel as rain screen

Fig. 175

edges of sheet flanged

rounded angles

external face

aluminium sheet 2.5mm thick deep drawn to form storey height window panel

flanged window surround

Single skin metal panel

Fig. 176

brackets as illustrated in Fig. 175. The flat sheet hangs from two or more supporting cleats, studs or pins near to the top edge of the sheet and is restrained by similar fixings that hold the sheet in position but do not restrain thermal movement. To be effective this arrangement requires accuracy in setting out and fixing to ensure adequate top support and reasonably close fitting restraint fixings that will not impede expansion and contraction of the sheet.

The flat sheet panels, illustrated in Fig. 175, are hung in front of the outer insulated metal panels as rain screens to provide an outer protection against wind, rain and solar heat. Flat sheet panels can be formed by drawing as window panels, as illustrated

in Fig. 176, with drawn flanged edges to the panel and around the window opening for stiffening and fixing. The deep drawn aluminium panels illustrated in Fig. 177 are screwed to separate outer aluminium carrier frames that are fixed to inner carrier frames through plastic thermal breaks. The aluminium carrier systems that support the single skin outer panels and the insulated panels behind are supported by a metal frame which is fixed back to the structure. Neoprene gaskets serve as weather and air seals. The complication of outer and inner carrier systems, thermal breaks, gaskets and insulated linings requires an accuracy in engineering skill that is not common to most building projects.

The single skin, deep drawn, profiled aluminium panels illustrated in Fig. 178 form spandrel rain screen panels. The panels are supported by lugs hung on pins fixed to lugs that are supported by the aluminium carrier system and the carrier system is similarly hung on lugs fixed to the structure so that there is freedom from restraint to allow for differential thermal and structural movements between the panels, the carrier system and the structure.

Rain screens

The term rain screen has been used to describe the use of an outer panel as a screen to an inner system of insulation and lining so arranged that there is a space between the screen and the outer lining for ventila-

167

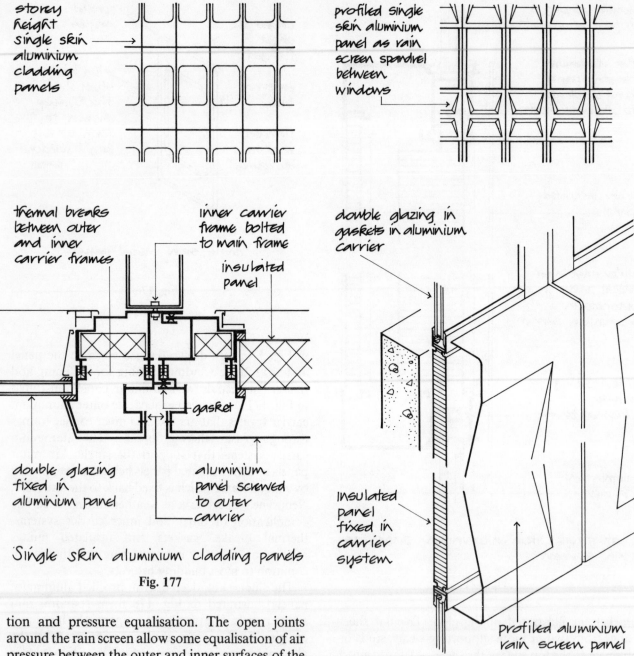

storey
height
single skin
aluminium
cladding
panels

profiled single
skin aluminium
panel as rain
screen spandrel
between
windows

thermal breaks
between outer
and inner
carrier frames

inner carrier
frame bolted
to main frame

insulated
panel

gasket

double glazing
fixed in
aluminium panel

aluminium
panel screwed
to outer
carrier

double glazing in
gaskets in aluminium
carrier

insulated
panel
fixed in
carrier
system

profiled aluminium
rain screen panel

Single skin aluminium cladding panels

Fig. 177

Profiled aluminium panel
as rain screen

Fig. 178

tion and pressure equalisation. The open joints around the rain screen allow some equalisation of air pressure between the outer and inner surfaces of the rain screen, which provides relief of pressure of wind driven rain on the joints of the outer lining behind the rain screen. Another advantage of the rain screen is that it will protect the outer lining system from excessive heating by solar radiation and protect gaskets from the hardening effect of direct sunlight.

All wall panel systems are vulnerable to penetration by rain which is blown by the considerable force of wind on the face of the building. The joints between smooth faced panel materials are most vulnerable from the sheets of rainwater that are

blown across impermeable surfaces. The concept of pressure equalisation is to provide some open joint or aperture that will allow wind pressure to act each side of the joint and so make it less vulnerable to wind driven rain. Plainly it is not possible to ensure complete pressure equalisation because of the variability of gusting winds that will cause unpredictable, irregular, rapid changes in pressure. Providing there is an adequate open joint or aperture there will be some appreciable degree of pressure equalisation which will reduce the pressure of wind-driven rain on the outer lining behind the rain screen. A fundamental part of the rain screen is air tight seals to the joints of the panels of the outer lining system to prevent wind pressure penetrating the lining.

Because of the unpredictable nature of wind-blown rain and the effect of the shape, size and groupings of buildings on wind, it is practice to provide limited air space compartments behind rain screens with limited openings to control air movements.

The single skin panels hung in front of insulated metal panels illustrated in Figs 175 and 178 serve as rain screens to the outer insulated metal panels and carrier systems.

Composite sheet metal panels

A thin sheet of metal provides negligible resistance to the transfer of heat and the thin sheet by itself has to be stiffened. The advantage of the combination of two thin skins of metal sheet with a core of insulating material is that the core provides insulation and the two skins some stiffness.

There are two forms of composite sheet metal panel, the box panel and the laminated panel. Box panels derive their stiffness from the two sheets of metal that are formed as shallow pans that are joined around an insulating core as illustrated in Fig. 179. Laminated sheet metal panels are formed from two skins of sheet metal that are glued to an insulating core, under pressure. The adhesion of the sheets of metal to the core provides a rigid panel that derives its stiffness from the sandwich of skins and core and that maintains its flat surface by virtue of the laminated form. Laminated panels are usually sealed with an edging of plastic or wood as illustrated in Fig. 180.

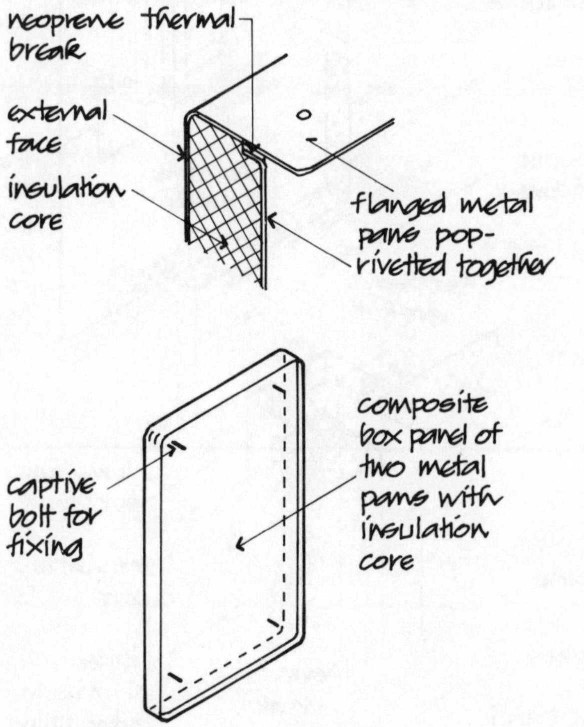

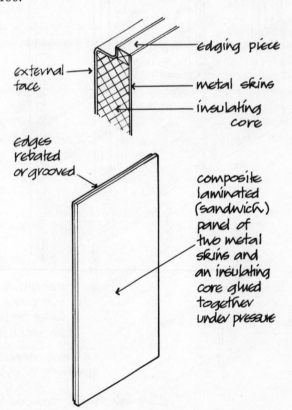

Composite metal box cladding panel

Fig. 179

Composite laminated (sandwich) metal cladding panel

Fig. 180

The disadvantage of composite panels is that the insulated core causes the outer skin of the panel to expand and contract more than the inner skin due to changes in air temperature and solar radiation so that the outer skin may bow out as it expands. This may cause apparent distortion of flat faced panels and delamination of laminated panels, which will lose some rigidity from loss of bond between the outer skin and the insulating core. To minimise distortion caused by delamination, the size of the panels should be limited, light or reflective colours should be used on the outer face to reduce solar heat gain and panels should not be rigidly fixed inside a carrier system that will restrain movement and so accentuate distortion of the outer skin.

The aluminium sheet, box panels illustrated in Fig. 181 were specifically designed for this building which is a notable example of the integration of the

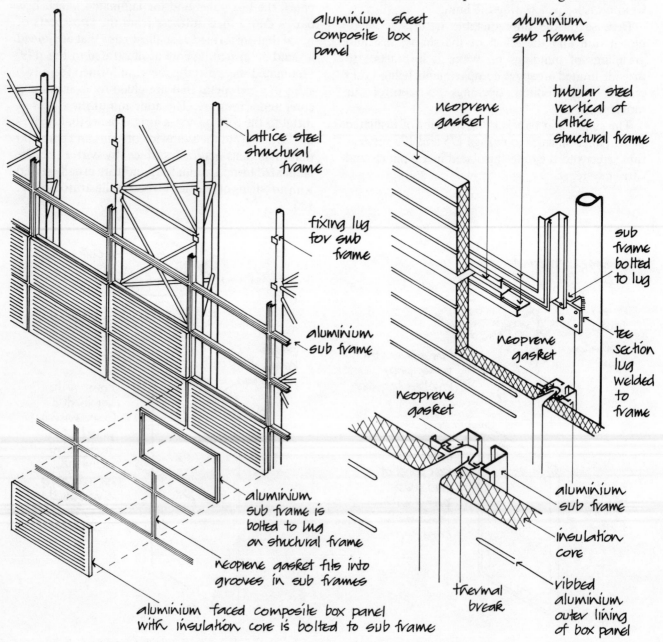

Sheet metal and insulation composite box panel cladding

Fig. 181

170

components of a building. The ribbed, anodised aluminium, box panels are made by vacuum forming. The outer tray is filled with phenelux foam and the inner skin then fitted and pop rivetted to the outer tray through a thermal break. Separate aluminium subframes for each panel are bolted to lugs on the structural frame. Continuous neoprene gaskets seal the open drained joints between panels which are screwed to subframes with stainless steel screws.

Jointing and fixing

Composite sheet metal panels have been in use as spandrel panels to curtain wall systems for many years and the jointing and fixing of these panels for use as external cladding has developed from the early use in curtain walling. The use of preformed gasket seals in aluminium carrier systems has developed with changes in curtain wall techniques so that today the majority of composite panels are fixed and sealed in neoprene gaskets fitted to aluminium carrier

systems fixed to the structural frame, as illustrated in Figs 177, 178 and 181, in which the carrier system supports the panels and the gaskets serve as a weather seal and accommodate differential thermal movements between the panels and the carrier system, whether the carrier system is exposed on the face of the building or hidden by open joints.

To reduce the effect of the thermal bridges made by the metal carrier at joints, systems of plastic thermal breaks and insulated cores to carrier frames are used.

Where open horizontal joints are used to emphasise the individual panels there is a flat or sloping horizontal surface at the top edge of each panel from which rain will drain down the face of panels and in a short time cause irregular and unsightly dirt stains, particularly around the top corners of the panels.

Single skin panels hung as rain screens in front of insulated panels will provide some protection from rain and wind and extreme changes of temperature and the hardening effect of sunlight on gaskets.

INDEX

Acrylic finish, 166
Admixtures, concrete, 77
Aerated concrete, 94
Aggregate
  artificial, 76
  coarse, 76
  fine, 76
  grading, 76
  natural, 76
  particle shape, 77
  surface texture, 77
Aggregates, 75
Alkali-silica reaction, 81
Aluminium sheets, 165
Anchorage of reinforcement, 82
Angle support, 132
Anodising, 166
Artificial aggregate, 76
Artificial stone, 125
ASR, concrete, 81
Assembling reinforcement, 85

Bars, deformed, 84
Base plates, column, 55
Bases, column, 55
Batching
  volume, 78
  weight, 78
Beam and filler block floor, 64
Beam and slab floor, 99
Beam and slab raft foundation, 14
Beam, castella, 38
Beams
  built-up, 54
  cantilevered, 83
Bearing pressure, 6
Bearing strength, 49
Bituminous membrane, 29
Block casing, 73
Board casing, 71
Bolt boxes, 57
Bolt pitch (spacing), 49
Bolts
  hexagon headed, 48
  high strength friction grip, 48
  turned and fitted, 48
Bond of reinforcement, 82
Bored piles, 15, 21
Boreholes, 3
Box frame construction 98

Box panel, metal, 169
Brick casing, 73
Brick cladding, 126
  restraint, 130
  support, 128
Built-up beams, 54
Bulb of pressure, 7
Bush hammering, 95
Butt weld, 52

Cantilever beam foundation, 12
Cantilever beams, concrete, 83
Carrier systems, 167
Cast stone
  cladding, 126
  facings, 131
Castella beam, 38
Cavity insulation, 126
Cellular raft foundation, 14
Cement, 74
  glass fibre reinforced, 149
  high alumina, 75
  low heat Portland, 75
  ordinary Portland, 74
  Portland blast furnace, 75
  rapid hardening Portland, 74
  sulphate resisting Portland, 74
  water repellent, 75
  white Portland, 75
Characteristics of aggregate, 76
Cladding
  brick, 126
  cast stone, 126
  natural stone, 126
  precast concrete, 137
Coarse aggregate, 76
Coarse grained soils, 5
Cohesion of particles, 4
Cohesive fine grained soils, 6
Cold bridge, 123, 162
Cold roll-formed sections, 39
Cold rolled steel deck, 64
Cold strip sections, connections, 58
Cold worked steel reinforcement, 84
Column base plates, 55
Column bases, 55
Column foundations, 55
Columns
  built-up, 54
  concrete, 86

Combined foundations, 11
Compacting concrete, 79
Composite construction, 112
Composite panels, GRC, 149
Composite sheet metal panels, 169
Composite sheets, 163
Compressibility, soils, 4
Compressible joints, 129
Compression seal, 162
Concrete, 74
Concrete admixtures, 77
    casing, 73
    cover, 82
    facings, 137
    frames, 96
    mixes, 77
    reinforcement, 81
    structural frames, 96
Concrete, aerated, 94
    ASR, 81
    compacting, 79
    construction joints, 79
    creep, 80
    curing, 79
    deformation of, 80
    foamed, 94
    lightweight, 93
    mixing, 79
    no fines, 93
    placing, 79
    prestressed, 90
    surface finishes, 94
    workability, 77
Connections
    cold strip sections, 58
    hollow sections, 58
    steel frame, 45
    welded, 54
Construction joints, concrete, 79
Contact pressure, 17
Corbels, 133
Cover beads, 162
Cover, concrete, 82
Cramps, 132
Creep, 80
Cross wall construction, 98
Curing concrete, 79
Curtain walling, 158

Deformation of concrete, 80
Deformed bars, 84
Design, methods of, 33
Designed mixes, 78
Differential settlement, 9
Displacement piles, 15
Double shear, 48
Dowel fixing, 134
Dowels, 134

Dragged finish, 95
Drained joint, open, 144
Driven cast-in-place piles, 17
Driven piles, 15
Drop slab floor, 100
Drying shrinkage, 80
Ductility, mild steel, 36
Durability, walls, 118

Elastic deformation, concrete, 80
Elasticity, mild steel, 36
End bearing piles, 19
EPDM, 162
Exclusion of wind and rain, 118, 162
Expansion joints, 129
Exposed aggregate finish, 95
External walls, 124

Face fixing, stone, 135
Facings, 131
    cast stone, 131
    concrete, 137
    faience slabwork, 137
    fixing, 131
    mosaic, 137
    natural stone, 131
    support, 128, 130
    terra cotta, 137
    tile, 137
Factor of safety, 34
Falsework, 90
Fasteners, steel frame, 45
Fill, 3
Fillet weld, 52
Fine aggregate, 76
Fine grained soil, 6
Finish, exposed aggregate, 95
Fire protection
    block casing, 73
    board casing, 71
    brick casing, 73
    concrete casing, 73
    intumescent coating, 69
    mineral fibre coating, 69
    plaster and lath, 73
    spray coatings, 68
    structural steel, 68
    vermiculite/gypsum/cement, 69
Fire resistance, walls, 119
Fire safety, 60, 68, 119
Fixed end support, 82
Fixing
    for facings, 131
    GRP, 154
    reinforcement, 85
    restraint, 130
    sheet metal, 171

Flanged GRP, 153
Flat slab (plate) floor, 100
Float glass, 160
Floor
  beam and slab, 99
  construction, 99
  drop slab, 100
  flat slab (plate), 100
  steel deck and concrete, 64
  waffle grid, 100
Flowdrill jointing, 57
Foamed concrete, 94
Formwork, 90
Foundation(s), 1
  beam and slab raft, 14
  bolt boxes, 57
  cantilever beam, 12
  cellular raft, 13
  columns, 55
  combined, 11
  pad, 11
  pile, 14
  raft, 14
  solid slab raft, 14
  strip, 11
Frame
  skeleton, 41
  structural steel, 40
Freyssinet system, 92
Friction piles, 17
Frost heave, 5
Functional requirements
  foundations, 1
  walls, 118

Galvanised steel reinforcement, 85
Gasket joint, 150
Gaskets, 150
Gel coat, 153
Gifford-Udall-CCL system, 92
Girder, Vierendeel, 41
Glass, 160
Glass fibre reinforced cement, 146
Glass fibre reinforced polyester, 154
Glass wall
  restraint, 161
  support, 161
Glazed wall, 156
Glazing, patent, 157
Grading of aggregate, 76
GRC, 146
  composite panels, 149
  jointing, 150
  open drained joint, 151
  restraint fixings, 151
  ribbed, 148
  sandwich panel, 149
  single skin, 147

  stud frame, 148
  support, 151
Grillage foundation, 56
GRP, 152
  fixing, 154
  flanged, 153
  jointing, 153
  profiled, 153
  sandwich panel, 154
  stiffening ribs, 153
  support, 154

Heat reflective glass, 160
Hexagon headed black bolts, 48
High alumina cement, 75
High strength friction grip bolts, 48
Hollow box mullion, 157
Hollow clay block floor, 104
Hollow floor units, precast, 102
Hollow sections
  connections, 57
  steel, 39

In-fill panels, 155
In situ cast concrete frames, 96
Insulation, cavity, 126
Internal friction, soils, 5
Intumescent coatings, 69
Inverted 'T' beam, 113

Jacked piles, 20
Joint, spigot, 161
Jointing
  GRC, 150
  GRP, 153
  sheet metal, 171
Joints,
  compression, 129
  construction, concrete, 79
  expansion, 129
  gasket, 150
  movement, 136
  precast concrete, 143
  sealed, 144
  stone, 136

Laminated panel, metal, 169
Laying up, GRP, 152
Lee-McCall system, 92
Lift slab construction, 110
Lightweight concrete, 93
Limit state method of design, 35
Load factor method of design, 34
Loadbearing fixings, 130, 132
Low heat Portland cement, 75

Made-up ground, 6
Magnel-Blaton system, 93

Manual metal-arc welding, 50
Mastic asphalt tanking, 27
Membranes, bituminous, 29
Metal inert-gas welding, 51
Metal support fixings, 131
Methods of design, 33
MIG welding, 51
Mild steel, 35
Mild steel reinforcement, 81
Mineral fibre boards and batts, 71
Mineral fibre coatings, 69
Mixes
  concrete, 77
  designed, 78
  nominal, 78
  prescribed, 78
  standard, 78
Mixing concrete, 79
MMA welding, 50
Moist earth method, 126
Mosaic, 137
Movement joints, 136
Movement of structural frames, 117
Mullions, hollow box, 157

Natural aggregates, 76
Natural stone cladding, 126
Neoprene, 136
No fines concrete, 93
Nominal mixes, 78
Non-cohesive soils, 5
Non-displacement piles, 15

Open drained joint, 144
Ordinary Portland cement, 74
Organic soils, 6

Pad foundations, 11
Panels
  in-fill, 155
  single skin, 162
Parallel beam frame, 59
Particle shape, aggregate, 77
Patent glazing, 157
Peat, 6
Permeability, soils, 5
Permissible stress method of design, 33
Pile cap, 24
Pile foundations, 14
Piles
  bored, 21
  displacement, 15
  driven, 17
  driven cast-in-place, 17
  end bearing, 17
  friction, 22
  jacked, 20
  non-displacement, 15

spacing, 24
Pin jointed frame, 44
Pitch (spacing), bolt, 49
Placing concrete, 79
Plain concrete finishes, 94
Plaster and lath, 73
Plasterboard casings, 72
Plasticity, soils, 4
Plate floor, 100
Point tooling, 95
Polyester powder finish, 164
Portland blast furnace cement, 75
Portland cement, 74
Post-tensioning, 93
Pre-tensioning, 91
Precast beam and filler floor, 104
Precast concrete
  cladding, 137
  plank floor, 102
  restraint fixings, 139
  support, 139
  tee beam, 103
  wall frames, 107
Precast hollow floor units, 102
Precast reinforced concrete frames, 104
Preflex beams, 113
Preformed casing, 72
Pressure bulb, 7
Pressure, contact, 7
Pressure equalisation, 167
Prestressed concrete, 90
Profiled GRP, 153
Profiled sheets, single skin, 163
PSC one-wire system, 93

Raft foundations, 14
Rain screen, 167
Rapid hardening Portland cement, 74
Rebated joint, 144
Rectangular grid frame, 42
Reinforced concrete floors, 99
Reinforced concrete frame, precast, 104
Reinforcement, 81
  anchorage, 82
  assembly, 85
  bond of, 82
  cold worked steel, 84
  deformed bars, 84
  fixing, 85
  galvanised steel, 85
  mild steel, 84
  spacers, 87
  stainless steel, 85
  stirrups, 85
Relative settlement, 9
Resistance to corrosion, steel, 36
Resistance to sound, 124
Restraint

brick cladding, 130
glass wall, 161
Restraint fixings, 130, 134, 151
Ribbed GRC, 148
Ribs GRP, 153
Rocks, 3

SA welding, 52
Sandwich panel, GRC, 149
Seal, compression, 162
Sealed joints, 144
Settlement, 9
Shear, 82
double, 48
single, 48
Shear studs, 113
Sheet metal
composite panels, 169
fixing, 166
jointing, 166
wall cladding, 166
wall panels, 165
Sheets
aluminium, 166
composite, 166
stainless steel, 165
steel, 165
Shrinkage, drying, 80
Single skin GRC, 147
Single skin panel, 166
Single skin profiled sheet, 163
Site exploration, 3
Skeleton frame, 41
Slimfloor construction, 66
Soils
cohesive, 6
non-cohesive, 5
organic, 6
Solar control glass, 160
Solid slab raft foundation, 14
Solid walls, 125
Spacers for reinforcement, 87
Spacing of piles, 24
Spigot joint, 161
Spray coatings, 68
Stainless steel reinforcement, 85
Stainless steel sheet, 165
Standard mixes, 78
Standard rolled steel sections, 38
Steel
cold roll-formed sections, 40
fire protection, 36
hollow sections, 39
mild, 35
weathering, 36
Steel deck and concrete floor, 64
Steel frame connections, 45
Steel frame fasteners, 45

Steel frame
pin jointed, 44
rectangular grid, 42
Steel grillage foundation, 56
Steel sections, 35
Steel sheets, 165
Steel tubes, 39
Stirrups, reinforcement, 85
Stone
face fixing, 135
joints, 136
Strength, mild steel, 36
Strength and stability, walls, 118
Strength of bolted connections, 48
Strip foundations, 11
Structural frames, concrete, 96
Structural steel frames, 40
Structural steelwork, fire protection, 68
Stud frame, GRC, 148
Submerged arc welding, 52
Substructures, 24
Sulphate resisting Portland cement, 74
Support angles, 132
Support
brick cladding, 128
fixed end, 82
glass wall, 161
GRC, 151
GRP, 154
precast concrete, 139
sheet metal, 163
Support for facings, 132
Surface finishes, concrete, 141
Surface texture, aggregate, 77

Tanking, 27
Tee beams, precast concrete, 103
Terra cotta, 137
Thermal bridge, 123, 162
Thermal properties, walls, 121
Tiles, 137
Tooled surface finishes, 94
Top hat section beam, 64
Toughened glass, 160
Trial pits, 3
Tubes, steel, 39
Turned and fitted bolts, 48
Types of aggregate, 76

Upstand beams, 98

Vermiculite/gypsum boards, 72
Vermiculite/gypsum/cement coatings, 69
Vierendeel girder, 41
Volume batching, 78

Waffle grid floor, 100
Wall

curtain, 158
glazed, 158
Wall cladding, sheet metal, 166
Wall frames, precast concrete, 107
Wall panels, sheet metal, 165
Walls
brick, 126
external, 125
functional requirements, 118
solid, 125
Water/cement ratio, 77
Water reducing admixtures, 77
Water repellent cement, 75
Waterstops, 25, 26
Weathering steel, 36

Weight batching, 78
Weld
butt, 52
fillet, 52
Welded connections, 54
Welding, 49
MIG, 51
MMA, 50
SA, 52
White Portland cement, 75
Wind bracing
concrete frames, 99
steel frames, 41
Workability, concrete, 77